Project Management
Concepts and Definitions

More than 1800 PM Concepts and
Definitions Explained
with 400 Examples

Dr. Muhammad Odeh

info@innovatepmsolutions.com
www.innovatepmsolutions.com

Library of Congress Control Number: 2020905266

ISBN: 978-1-7338296-4-9

Printed in the United States of America
First printing edition 2020.

About the Author

Dr. Muhammad Odeh

- Author of two Books "Transforming Project Management" In the age of the Internet of Things and Artificial Intelligence. And Navigating The PMP® A PMP Study and Review Guide.
- 25+ years' experience in IT Management, Strategy, Direction, Architecture, and Project Management.
- Doctor of Engineering in Engineering Management. The George Washington University, College of Engineering and Applied Science, Washington DC, USA.
- Master of Science in Computer and Decision Science, Pratt Institute, Brooklyn, New York, USA.
- Bachelor of Science in Computer Science, Missouri State University, Springfield, Missouri., USA.
- Stanford University Advanced Project Management (SAPM), Stanford University, San Francisco, California, USA.
- Business Leadership Certificate, Ashridge Business University, Ashridge, UK.
- Sustaining High-Performance Certificate, IMD, Lausanne, Switzerland.
- Certified in Risk and Information Systems Control (CRISC), ISACA, USA.
- Certified Information Security Manager (CISM), ISACA, USA.
- Information Technology Infrastructure Library Certification (ITIL).
- Certified Business Continuity Professional (CBCP), DRI, USA.
- Project Management Professional (PMP), PMI, USA.
- Master Certificate in Project Management. The George Washington University, Washington DC, USA.
- Computer Systems Security Professional (CSSP) - Senior Grade, IACSS, USA.

Dedication

To my late father Fuad Odeh whose relentless passion for education was and in so many ways continues to be a beacon of light, his searing and profound 50 years commitment to education was unprecedented, and vision thereof was so vivid in color that I could not classify along a traditional spectrum. To the purity and beauty of the soul in my mother, the beloved Majed, Mukhles, Musab, Maysoon, May, Manal, Maisa, and Maram.

Most importantly, to my wife Ansam and our children Fuad, Ziad, and Farah, the source of my joy and inspiration, for their continuous assurance and encouragement to complete this work and helping me comprehend with keen awareness the real depth of affection.

The sun rises and sets with them.

Introduction

My first book, "Transforming Project Management – In the Age of the Internet of Things and Artificial Intelligence," is a work of experience, practice, and research on contemporary issues in project management. I examined why the mechanism invented to improve project management has proved abortive or ineffective, the factors that trigger the failure of these techniques to improve project management success, and how the advantage of turning innovative technologies into tools and make diligent use of it will increase the efficiency of project management. The book introduced Project Management Transformation Framework (PMTF) and Innovation Project Management Office (IPMO). The PMTF presents a framework that uses emerging technologies and information intelligence, which will enhance the chances of project success.

My second book, "Navigating the PMP." A Project Management Professional Exam study and review guide. The book is intended to help readers pass the PMP exam with chapters addressing all the knowledge areas, each chapter includes key focus areas, sample questions, and concludes with 20 practice questions on each area. The final chapter of the book contains 100 additional practice questions.

This book goes back to basics; it is meant to explain most of project management glossaries and terms and provide examples where needed to illustrate the meaning. The book includes more than 1800 project management related concepts and terms, along with more than 400 examples.

1

A

Acceptable quality level: A statistical tool to inspect a particular sample size.

Acceptance criteria: A project performance requirements or essential conditions that must be achieved before a delivery is accepted (For example, an acceptance document or template that outlines all the criteria and measures for a project or a milestone to be signed off and accepted).

Acceptance date: The date on which the project owner or sponsor formally provides a final acceptance of the deliverable item or project.

Acceptance review: A process by which a product or service presented for acceptance complies with its agreed specifications and functionality (For example, a document or template that outlines the method and procedures by which acceptance is reviewed for approval).

Acceptance sampling: A quality control statistical procedure used to test a sample of units to determine if the sample complies or exceeds a given percentage (For example, a document or a template specifying the number of units or samples to be tested for acceptance).

Acceptance test procedure: A formal, predefined documented set of instructions used to prepare the acceptance test and evaluation of the test results to determine the compliance of the deliverables with the acceptance criteria.

Acceptance testing: A formal application of performance measurements to ensure project deliverables comply with agreed specifications and functionalities.

Accountability matrix: See Responsibility assignment matrix.

Accountability: The responsibility or obligation of a single resource or a functional organization for the completion of an activity.

Accounting period: A pre-determined period in which project revenues and costs are posted for analysis (For example, a week, a month, or a quarter).

Accounts payable (AP): A list of project liabilities based on purchase or money owed by the project to its suppliers (For example, services, equipment, or supplies).

Accounts receivable (AR): A list of monies due on current accounts based on the sale of products or services. The balance of funds due to the project for goods or services delivered.

Accreditation: A process or an act of granting recognition in which certification of competency, authority, or credibility is presented.

Accrued cost: A project cost of goods or services reserved for which payment is due but has not been made.

Acid test: A conclusive and thorough examination of the success of a deliverable or a project (For example, a rigorous and severe reliability test).

Acquisition control: A systematic and uniform process of managing an acquisition. (For example, a consistent way of acquiring project services and equipment).

Acquisition methods: A method by which products or services are acquired. (For example, requiring bidders to submit financial or technical proposals in a certain way).

Acquisition plan: A formal document that involves all aspects of an acquisition. (For example, acquisition technical, and financial management).

Acquisition process: A formal documentation of the process undertaken by the performing organization that involves acquisition. (For example, the process of acquiring products or services for the project or parts of it).

Acquisition strategy: A formal document involves the acquisition policy and approach (For example, the plan of acquiring products or services for the project or parts of it).

Acquisition: Acquiring products or services for the project or for the use of the delivery organization (For example, purchasing or leasing products or equipment for the project or parts of it.)

Action item: An activity agreed to be completed by a resource or a functional organization (For example, a task assigned to a resource or afunctional organization as a result of assignment matrix or risk response strategy).

Action plan: A formal document involves the details of project activities and tasks with their assigned resources (For example, what needs to be done, by whom, and by when).

Active listening: A skill that involves paying close attention with all senses to what is being said (For example, listening carefully with full concentration to what is being said and asking for clarification in case of any ambiguity rather than just passively hearing the message.)

Active: A project status describing a project with assigned resources and activities.

Activities ID: A unique code identifying each activity in a project (For example, identification of specific activities to be performed to achieve project results).

Activity-based costing: A costing method that identifies activities in a project and assigns the cost of each activity to all products or services based on their resource consumption (For example, an indirect or overhead cost is attached to a service).

Activity-based management (ABM): A method of identifying and evaluating activities that a project performs using activity-based costing to improve operational decisions (For example, activity-based indirect and overhead costs are used to be traced to products, services or projects).

Activity definition: Identifying the specific activities to be performed to achieve the project results (For example, a list of all activities that need to be completed for the project to achieve its desired results).

Activity description: A label or short description used in a project network diagram (For example, a brief description of the scope of work of the activity).

Activity duration estimating: An estimate of the required number of work periods needed to complete specific activities (For example, activity A will need four days to complete).

Activity duration: An estimate specifying the length of time needed to complete a project activity (Form example, number of hours, days, or weeks required to complete an activity taking into consideration the nature of the activity and resources required for it).

Activity file: A file that includes details of all project activities (For example, activity ID, definition, description duration).

Activity list: A formal numbering scheme of project activities to be completed that is based on the project's WBS (For example, activities organized as 1 then 1.1 then 1.1.1, etc. to include all and only activities and sub-activities of the project to complete).

Activity on arc: See Precedence diagramming method.

Activity on arrow (AOA): See arrow diagramming method.

Activity on arrow network: Arrow diagram, network in which the arrows represent the activities.

Activity on node (AON): See Precedence diagramming method.

Activity on node network: Precedence diagram, a network in which the nodes represent the activities.

Activity sequencing: A systematic, documented way of identifying activities and their dependencies.

Activity status: An activity state of completion (For example, an activity is complete, pending, scheduled).

Activity: An element of work to be completed during the life of a project (For example, an activity or a task must have a start and finish date, dependencies, duration, resources).

Actual cost (AC): Actual amount made to acquire products, services, or assets (For example, the actual cost incurred to provide a service that could be different from forecast or market cost).

Actual cost of work performed (ACWP): Total direct and indirect costs incurred in completing work during a given period. See also earned value analysis.

Actual dates: Actual dates reflect real and actual dates that project activities started and finished, actual dates are recorded as the project progresses (For Example, exact dates are recorded once reached as opposed to planned or forecast dates).

Actual finish date (AF): Actual date that project activity is completed.

Actual rate: Actual expenditure as they are entered in the project's accounting system (For example, actual amount, fee or price.)

Actual start date (AS): Actual date that a project activity started.

Actuals: Actual cost or revenue incurred in the performance of project activity (For example, the actual amount gained, or the actual cost incurred as a result of completing a project, a milestone, or an activity).

ACWP: Actual cost of work performed.

Adjourning: The last phase of team building where the team disbands or breaks up.

ADM: Arrow diagramming method

Administrative closure: An arranged set of activities involving a formal project completion (For example, generating, collecting, and distributing information to formalize acceptance by the project by the sponsor, client, or customer).

Administrative expense: Expenditure that is usually allocated to a project as a result of joint administrative expense (For example, shared office space between multiple projects, the total cost is spread among the involved projects per square meter or foot usage of the space).

Advance payment: A payment that is made in advance in anticipation of agreed performance or completion of project activities. (For example, advanced payment authorized by the project manager to one of the project's subcontractors).

Advanced material release: A documented authorization to initiate the purchased material that is expected to have long lead-time or time-critical materials (For example, a project manager authorizes the releases of a purchase order for material that takes a long time to be available).

AF: Actual finish date

Affiliates: Direct or indirect project-related individuals or organizations.

Agreement: A formally negotiated, typically legally binding agreement between two or more parties

Allocated baseline: A performance baseline allocated to a function of the product.

Allocated cost: A type of expense that is charged or assigned to a defined business process (For example, allocated cost could include sales cost, fixed cost).

Allowable cost: Expenses that are included in a contracted product or service and are recovered in contracts between two companies.

Alternatives analysis: Analysis and evaluation of different choices to achieve an objective (For example, providing an approach to a problem that may include workarounds or trade-offs, comparison of different factors like operational cost, risks).

Alternatives identification: A planning process of identifying different approaches and techniques to solve completing the project (For example, implemented a change request to update your project).

Ambiguity: Contract language or terms that could be understood to have the quality of being open to more than one interpretation.

Ambiguous jurisdictions: A concealed conflict or conflict source in organizations when the organization has not clearly defined individual areas of decision authority (For example, conflict occurs when two resources or functional groups believe they have the responsibility for the same activity).

Amendment: A minor change or addition to a contract or legal document.

Amortization: A process of gradually writing off the initial cost of an asset (For example, dropping or settling a debt with regular payments.)

Amount at stake: The impact of positive or negative consequences that could occur to a project if a specific risk occurs (For example, the amount of money a project stands to lose if a particular risk occurs).

Analogous estimating: An estimate using past or present similar projects or activities as a base for estimating activity duration or cost (For example, expert judgment and top- down estimating are also referred to as analogous estimating).

Analytical approach: Analysis approach to breaking down a problem into the elements necessary to solve it (For example, use of formal analysis tools to break a problem down to its smallest possible parts to solve it).

AND relationship: A logical connection between two or more activities that join or deviate from an event (For example, every activity joining an event must be performed).

Annual basis: A reference to a return earned throughout 12 months (For example, a statistical technique in which projections of less than a year's worth of data to project a full year's worth of returns).

Annual percentage rate (APR):   The annual percentage rate charged for borrowing or earned through an investment (For example, Cost of credit that to be paid over 12 months).

Annual receipts: The average gross revenue of a project or a subproject.

Annual report: A formal financial statement that includes descriptions of the project's operations for the year.

Anticipatory breach: A preemptive break of a contract before a performance assessment is due (For example, a contractual party expresses an intention to break the contract by not performing (or otherwise acts in such a way to the conclusion that the party does not intend to achieve).

AOA: Activity-on-arrow

AON: Activity-on-node

APIs: Application programming interface.

Apparent low bidder: A potential contractor who has submitted the lowest compliant bid for a project as described in the request for proposal.

Application area: A class of projects with commonalities that maybe not shared in all projects (For example, construction project vs. manufacturing projects or government project vs. none government projects).

Application prototype: A working and interactive model of the product or solution.

Application: A reference to the use of specific tools and techniques in a project (For example, word processing, spreadsheets, or task scheduling software).

Applied cost: An accounting term that refers to an assigned cost to an object, which may be different from the actual cost (For example, the applied cost is determined for each cost object using an allocated cost, also see allocated cost).

Applied rates: Rates applied to projects by the organization for budgeting and reporting purposes (For example, project managers account for these rates as part of their budgets and usually set them above expected rates).

Apportioned cost: A cost item that cannot be directly charged to a cost center, such an item is prorated amongst various cost centers

Apportioned effort: A technique of planning and measuring the earned value of an effort that is not readily measurable. Or broken into discrete work packages (For example, Inspection is a typical use of this method).

Approach statement: How the management of a project is carried out and the way the project will accomplish its intended objectives.

Approval to proceed: An endorsement granted a project at initiation or before the beginning of the next stage.

Approval: A term used when an accepting party determines that a deliverable or a project as fit for purpose.

Approved bidders list: An approved list of prequalified contractors to submit proposals.

Approved change: A change to the project scope, schedule, or budget, which has been approved for implementation.

Approved: A status describing a consent before the start of project activities or the next phase of a project.

AQL: See Acceptable quality level.

Arbitration: A legal technique for the resolution of a dispute outside the court (For example, a bargaining agreement between two parties to settle a dispute outside the court).

Arbitrator: An independent person or official appointed to settle a dispute where the disputed parties agree to accept the arbitrator's decision (For example, an impartial person appointed to resolve a disagreement between two parties.)

Arc: A line that is connecting two nodes.

Arrow diagram: See Activity on arrow network.

Arrow diagramming method (ADM): A network diagramming method where activities are represented by arrows, the start of the activity is the tail of the arrow, and the finish of the activity is the head of the arrow. Activities are connected at points called nodes to show the sequence in which the activities are to be performed. (the extent of the arrow does not necessarily represent the expected duration of the activity). See also Precedence diagramming method

Arrow: A graphic presentation of an activity used in ADM. The tail of the arrow denotes the start of the activity, and the finish of the activity is denoted by the head of the arrow. The length of the arrow has no relation to the duration of the activity.

As-built schedule: A project schedule representing actual start, duration, and finish dates (For example, as-performed schedule).

As late as possible (ALAP): An activity where the early start date is set as late as possible with no impact on the early dates of a successor activity.

As of date: See Data date.

As performed schedule: See As-built schedule.

As soon as possible (ASAP): An activity for which the early start date is set to be as soon as possible. (For example, ASAP is the default activity type in most project management systems).

AS: Actual start date

Asset: An item of property owned by a person, a project, or an organization regarded as having value (For example, cash accounts receivable, investments, notes receivable).

Assignment of contract: A transfer of contract obligations by one party to another party.

Associated revenue: Project charge that is of a revenue nature, therefore it is charged as incurred to the profit and loss account.

Assumption: Something taken or considered to be accurate, real, or specific. Assumptions generally involve a degree of risk.

Audit: A systematic examination of the project, or part of it to measure conformance with determined values (For example, financial audit, quality audit, design audit).

Authority: The power a person has to get other people to perform or act based on his/her decisions.

Authorization: The decision that triggers the assignment of resources or funding needed to perform activities or projects.

Authorized unpriced work: A scope change for which authorization to proceed has been approved; however, the estimated costs of the change are not yet determined.

Authorized work: A defined effort for which authorization has been approved.

Autocratic management style: A management style where one person controls all the decisions and takes minimal inputs from other group members (For example, a project manager makes decisions based on his/her own beliefs and does not involve others for their suggestion or advice).

Automatic decision event: An event that depends only on the outcome of the other activities, and that can be programmed or made automatic.

Award agreement: A written agreement between an organization and a participant evidencing the terms and conditions of an Award.

B

BAC: Budget at completion

Backward pass: The calculation of activity late finish dates and late start dates for the part of project schedule activities that have not been completed. This is determined by starting at the project's scheduled end date and working backward through the schedule network logic, and the project sponsor may set the end date. See also Network analysis.

Backward scheduling: Planning the tasks from the project task due date to determine the start date and any changes required.

Balanced matrix: An organizational structure where functions and projects have the same priority.

Balanced scorecard: A measurement technique that enables organizations to clarify their objectives and translate them into action to improve performance and results continuously.

Bar Chart: A chart showing a schedule of project activities and milestones. Activities are typically listed with other tabular information along the left side of the chart with a timeline across the top or bottom. Also called a Gantt chart.

Baseline cost: An estimated amount of money an activity was intended to cost when the schedule was baselined.

Baseline dates: An initially planned start date and finish date for all project activities (For example, baseline dates are used for comparison of planned dates and actual dates to calculate any deviation and as an earned value analysis tool to calculate budgeted cost of work scheduled).

Baseline finish date: See Scheduled finish date.

Baseline review: A review that involves walking through the project activities to verify that the project complies with the performance measurement baseline (For example, a baseline review will assess the accuracy of the related resources, budgets, and schedules and identify potential risks).

Baseline schedule: A baseline schedule is a fixed project schedule and the standard by which project performance is measured.

Baseline Start Date: See Scheduled start date.

Baseline Survey: A pre-project status information of participant conditions in which performance indicators will be compared against at midpoint and the end of the project.

Baseline: A point of reference, a plan used as the comparison point for project control reporting (For example, three baselines are used in a project, schedule baseline, cost baseline, and scope baseline where the combination of these baselines is referred to as the performance measurement baseline.)

BCWP: Budgeted cost of work performed

BCWS: Budgeted cost of work scheduled

Benchmarking: An evaluation of product or service by comparison with a standard. See also Competitive benchmarking.

Beneficiary: A person or an organization that is the primary beneficiary of the project (For example, the primary beneficiary is the main stakeholder that has a significant authority regarding the acceptance of the project).

Benefit-cost analysis: A process of direct and indirect cost estimation of project alternatives (For example, deciding the best alternatives based on measuring financial return on investment and payback periods).

Benefit-cost ratio (BCR): A financial ratio used in project selection by dividing the benefit of the project by its cost (For example, a quantified benefit of $1000 vs. a quantified cost of $800. A BCR more significant than one shows a financially profitable project, and less than one shows a financially losing project, in this case, 1000/800 is more than one).

Benefit measurement method: A method of project selection based on the present value of the investment and revenue generated by the project (For example, the cost and benefits of each project are calculated and compared to decide on the project that provides the highest benefits).

Benefits framework: A documented outline of the expected benefits of the project, the business operations involved, and performance measures.

Benefits management plan: A documented plan describing who would be responsible for achieving the project's benefits and how these benefits would be achieved, measured, managed, and monitored.

Benefits Management: An organizational process of benefits planning, identification, monitoring, and realization of those benefits.

Benefits: A reference to the efficiency, economy, and effectiveness of a future business or other operations to be delivered by a project.

Best and final offer (BAFO): A final technical and financial offer to perform or carry out project activities following negotiation that incorporates all agreed changes and modifications.

Best efforts: A type of contractual obligation that involves a best attempt to meet an objective (For example, a contractor's use of their best efforts to perform project activities within the planned budget and schedule).

Best practice: A result of learned experience and lessons learned on similar projects.

Beta test: A test of a product or a solution in a simulated environment where the results may be used for the intended application.

Biased sample: A sample that is not random and does not truly represent the total.

Bid analysis: An analysis of bids or tenders.

Bid bond: A legally binding instrument issued as part of a product or service by the contractor to the project owner to provide a guarantee. Intended to guarantee that the winning bidder will undertake the contract under the terms at which they bid

Bid guarantee: A form of security intended to ensure that the bidder will not withdraw from a bid within the time specified for acceptance and will execute a written contract (For example, bidder will provide required bonds including any necessary coinsurance or reinsurance agreements, within the time specified in the bid.

Bid list: A list of contractors invited to submit bids for products or services.

Bid opening: A public opening of sealed bids submitted by invited bidders.

Bid protest: A process where an unsuccessful contractor may challenge a possibly unjust contract award.

Bid/no-bid: An approach an organization may take to go ahead or not with submitting a proposal in response to a request or a tender invitation.

Bid: A tender, quotation, or an offer to enter into a contract to perform project activities described in the request for proposal or tender documentation.

Bidder: The entity or the organization that submits a bid.

Bidders conference: A meeting set by the project owner with potential contractors before preparation of proposals (For example, a meeting is a setup by the project owner to clarify and answer questions, make sure that all potential bidders have a common understanding of the procurement process).

Bidding period: The time between the date of the bid invitation and the date set for receipt of bid proposals.

Bilateral contract: An agreement between two parties in which each side agrees to fulfill their side of the contract.

Bill of lading: A detailed list showing receipt of goods for shipment issued by the organization in the business of transportation to the party responsible for consigning the goods (For example, a list of goods for shipment given by the master of the ship to the person responsible for consigning).

Bill of materials (BOM): A list of physical components required to build a project or a solution (For example, a list of material, assemblies, subassemblies, and elements required to assemble an office complex).

Bill of quantities (BOQ): A document that provides project-specific quantities of the items identified by the specifications in the tender documentation.

Binding arbitration: Means that the disputing parties must adhere to the arbitrator's decision and usually cannot appeal the decision to a court. See also Arbitration.

Blue teams: In cybersecurity projects, a blue team is defensive security professionals responsible for maintaining internal network defenses against all cyber-attacks and threats. See Red team

Body language: A nonverbal communication, in the form of facial expressions, physical posture, or gestures that either complements or contradicts the spoken word.

Boilerplate: A reference to contract clauses and terminology that are deemed standard but essential in a contract.

BOM: See Bill of materials.

Bond: A legally binding instrument issued as part of product or services supply bidding process, to provide a guarantee, that the winning bidder will undertake the contract under the terms at which they bid

Bonus: An additional payment or something of value paid to a resource or contractor for superior performance.

Bottleneck: A constraint that restricts performance, information flow, or a process.

Bottom-up estimate: A method of making estimates for every activity in the WBS and summarizing them to provide a total project cost estimate.

Brainstorming: An unstructured problem-solving technique for generating ideas by a project team and other stakeholders for planning purposes, risk identification, and other project-related activities.

Breach of contract: A legal cause of action in which a binding agreement is not honored by one or more of the contractual parties to the contract by non-performance or interference with the other party's performance.

Breach: Failure to carry out a contractual commitment. See also Material breach.

Breakdown structure: A hierarchical structure involving project elements being broken down or decomposed. See also work breakdown structure (WBS), product breakdown structure (PBS), and organizational breakdown structure (OBS).

Break-even chart: A graphic representation that is plotted on time or volume scale showing the relationship between total value earned and total costs.

Break-even point: A point in time during project execution, at which value earned equals total cost.

Budget at completion (BAC): The estimated full cost of the project at the time when it is complete.

Budget cost: The anticipated cost at the start of a project, a conversion of estimate into unit rates (For example, budget cost is used to compare actual cost and planned cost to determine variances and evaluate performance).

Budget decrement: The amount of a reduction in available activity funds.

Budget element: An item on the budget that has an allocated cost or revenue (For example, resources: the people, materials, or other entities needed to perform project activities).

Budget estimate: An estimate prepared in the early stages of a project to establish financial viability or secure resources.

Budget monitoring: Measuring actual project cost against the project budget to establish the variance (For example, analyze variances, and take the necessary corrective action).

Budget unit: A user-defined budget unit used for the calculation (For example, unit of hours. US Dollars, meters, or kilograms).

Budget update: A change to an approved cost baseline.

Budget: An amount allocated for the project that represents the estimate of planned revenue and cost. (For example, a budget may be expressed in terms of money or resource units (effort).

Budgetary control: A systematic approach to creating budgets, monitoring progress, and taking appropriate action to achieve budgeted performance.

Budgeted cost of work performed (BCWP): The budget for activities completed during a period (For example, BCWP can be calculated by taking the percentage of work completed times the baseline cost of the activity).

Budgeted cost of work scheduled (BCWS): The budget for activities scheduled to be completed during a period (For example, the planned cost of work that should have been completed according to the project baseline dates).

Budgeting and cost management: An estimate of costs and the agreed budget, and the management of actual and forecast costs against that budget.

Budgeting: A time-phased financial requirements.

Bulk material: Material organized in lots (For example, material purchased in quantity, weight where no specific item is different from any other in the lot).

Burden: Overhead expenses distributed over direct labor and/or material base. See also Indirect cost

Bureaucratic authority: A type of influence exercised by someone with knowledge of how the organization works (For example, someone who is very familiar with the organization's procedures, rules, and regulations).

Burn rate: A rate at which a project is losing money, rate of project expenditure per defined period (for example, total dollars per day or week).

Business agreement: A commitment to perform project activity or a group of activities leading to a deliverable or completion of a project.

Business case: Essential information needed to drive approval, authorization, or assess a project proposal and reach a reasoned decision. See also Project business case.

Business process model: A process of decomposing business process or functional areas (For example, the model shows a graphical representation of a functional area broken down into processes then subprocesses; and so, on down to down into activities).

Business process reengineering: A method of examining and redesigning business processes and workflows to drive better performance.

Business risk: An exposure a project has to factor that will lower its profits or lead it to fail (For example, a significant risk threatens a company's ability to meet its objectives or achieve its financial goals).

C

Calendar date: A specific date represented on the calendar (For example, July 3) as opposed to a relative date. See Relative dates.

Calendar start date: The first unit of the working calendar.

Calendar Unit: The lowest unit of time used in scheduling project activities (For example, hours, days, or weeks).

Canceled: Project status given to a canceled project.

Capability maturity model (CMM): A methodology used to develop and refine the relative maturity of an organization's software development process. The model describes a five-level evolutionary path of increasingly organized and systematically more mature processes.

Capability validation: A technical verification process of a system to assess its ability to satisfy a project's functional requirements.

Capacity assessment: An assessment analysis to measure the ability of the project to implement its strategy.

Capital assets: Physical property of any kind that is held by an organization (For example, all types of property tangible or intangible, fixed or circulating).

Capital cost: A balance sheet cost item that refers to acquiring an asset capable of performing its intended function over time. See also Revenue cost.

Capital employed: Amount of investment in a project that represents the sum of fixed and current assets, less current liabilities at a specific date.

Capital expenditure: Any money spent by a business or organization on acquiring or maintaining fixed assets (For example, land, buildings, and equipment).

Capital: Cash, equity, and debt and all other property of an organization.

Capitalization: A financial handling of asset expenditures compared to regular expenses (For example, organizations are required to capitalize costs of assets if they exceed a predefined amount in certain regions).

Career development: The process of managing education, work, and leisure to move toward a personally determined and evolving preferred future.

Cascade chart: A bar chart displaying vertical project activities on which the order of activities so that every project activity is dependent only on activities higher in the list.

Cash flow: Amount of cash receipts and payments in a specified period.

Cash-flow analysis: An examination of different project cash inflow and outflow within a predefined period (For example, cash in and out during June from different project activities Such as operating activities, investing activities, and financing activities).

Category of material: A specific group of material that consists of identical or interchangeable elements and is used during project performance (For example, plastic, wood, or ceramic).

Cause-and-effect diagram: See Ishikawa diagram.

CCB -Change Control Board

Cellular processing: A manufacturing process that is part of the just-in-time manufacturing encompassing group technology. The goal of it is to move as fast as possible, make a variety of similar products with little waste as possible.

Center of excellence (CoE): A team of skilled knowledge resources whose mission is to provide the organization with best practices around a particular area of interest. CoEs are often developed when there is a knowledge shortfall or skills gap within an organization.

Centralized contracting: A function within the organization or program whose responsibility is managing all aspects of the contracting process (For example, the person in charge of the function is accountable for contracting activities and performance).

CER: See Cost estimating relationship.

Certification: An accreditation by a neutral third party following an evaluation of a person or a system (For example, PMP or ISO 9000).

Champion: An end-user representative often part of a project team that acts as an advocate for a project.

Chance: A possibility of an outcome in a situation. See also probability.

Change control board (CCB): A formal assembly of stakeholders responsible for approving or rejecting changes to the project baselines.

Change control system: A formally documented procedure that defines how an official project may be changed.

Change control: A process that ensures potential changes to any baseline or deliverables of a project or any of its parts are recorded, evaluated, authorized, and managed.

Change in Scope: See Scope change.

Change log: A documented record of all project changes, proposed, approved, or rejected.

Change management plan: A documented approach to requesting, evaluating, approving, and implementing changes to the project or parts of it.

Change management: The formal process of how changes to the project are introduced, planned, and approved.

Change order: A formal documented instruction issued by the project owner requesting a change in the project (For example, a contractual change).

Change request: A formally documented request for a change to any aspect of the project (For example, a change to the schedule of certain project activities. The main difference between change order and change request is that a change order usually refers to a contractual change while a change request is mainly a change to the scope or schedule).

Change: A deviation from a planned value or event. The most significant changes in a project related to scope or schedule.

Changed conditions: A modification of the contractual agreement from the agreed and planned contract.

Charismatic authority: A type of authority influenced by an individual's personality (For example, someone is charismatic and liked by others may use that to influence other people).

Chart of accounts: A numbering system used to monitor project costs by category (For example, effort, goods, materials). The project chart of accounts resembles the organization chart of accounts of the performing organization. See also code of accounts).

Charter: A formal document established by senior management to sanction the project and authorize the project manager to carry it out within the scope, quality, time, cost, and resources. See also Project charter.

Checklist: A structured tool for a set of activities or steps to be performed. It helps to ensure consistency and completeness in carrying out a task (For example, to-do list or a list of identified risks).

Child activity: A sub-task or activity belonging to a parent task or activity existing at a higher level in the work breakdown structure.

Claim: An expression of the right to something that belongs to an entity (For example, a request for payment).

Clarification: A class of communication to remove minor ambiguities, irregularities, or miscommunication.

Classification of risks: A process used to identify risks to determine its type and priority, mitigation based on the risk evaluation process.

Clearance number: Refers to the number of successive units that passed an inspection successfully before a change is made to the inspection procedure.

Client or Customer: The main stakeholder, representing the user who commissions the project and would be paying for it.

Closeout: the completion of work on a project.

Closeout phase: The final phase in the project lifecycle where all administrative and financial items are completed and documented (For example, closeout phase includes documentation and archiving of project records and settling all financial transactions to be followed by documentation of lessons learner).

Closing processes: Project activities associated with formal acceptance of a project milestone or an entire project (For example, acceptance testing, and approval followed by a hand over to the end-users).

Closure: The formal endpoint of a project.

CMM: Capability maturity model

COCOMO: See constructive cost model.

Code of accounts: A coding system used to provide a unique identity of each element of the work breakdown structure. See also Chart of accounts.

Code of ethics: A formal statement of values and principles related to the behavior of the individuals employed in a project or an organization.

Coercive authority: A type of influence founded based on fear (For example, people do what is asked of them in fear of the consequences if they do not).

Collaboration software: Otherwise referred to as groupware is application software designed to help resources working on a standard task to attain their goals (For example, collaborative software such within a project creates a collaborative working environment).

Collective bargaining agreement: A formal agreement with a committee representing a group of people with a common interest.

Colocation: Location of project team members in the same physical location to enhance their performance and productivity.

Combative management style: A management approach where the project manager displays an eagerness to fight or be disagreeable over any given situation.

Command failure: A system failure caused by incorrect commands or signals from the operator or other components.

Commercial off the shelf (COTS): Refers to items, tools, software, or service available in the commercial market.

Commissioning: Progression of an installation from a static completion to full working order and achievement of the specified operational requirements.

Commitment: An obligation to fulfill a requirement (For example, A binding financial obligation in the form of a purchase order or contract).

Committed costs: Legally committed cost despite a delivery has taken place with invoices neither raised nor paid.

Commodity: Tangible goods or products.

Common cause variation: A fluctuation caused by unknown factors resulting in a random distribution. A measure of how well the process can perform when a unique cause variation is removed. See also Special cause variation.

Communicating: An active exchange or transfer of information from one party to another through a standard system of symbols, signs, or behavior. (For example, communication includes a message, a sender of a message, the receiver of a message and the medium used to transmit the message).

Communication barrier: An obstruction to active exchange or transfer of information (For example, barriers can be physical, cultural barrier, or logistical).

Communication blocker: Something which can create a divide between people and can impede the flow of effective communication (For example, very often people end up using communication blockers in their communication; sometimes intentionally and sometimes unintentionally).

Communication channel: Refers to the means and number of communication lines between project stakeholders. Means of communication refer to formal communication, informal communication, and unofficial interpersonal communication within the organization's social structure. At the same time, the number of lines depends on the number of stakeholders a project manager needs to communicate with (For example, team members, executive management and customers are all stakeholders).

Communication management plan: Part of the overall project management plan, a document that describes the processes, procedures, tools, and techniques for planning, gathering, distributing, and storing project information.

Communication management: The process of identifying the project stakeholder's communication needs and requirements and making sure that these requirements are appropriately addressed.

Communication model: A process that involves the different parts of communication such as the sender, receiver, the message or information to be communicated and the mode of communication.

Communication planning: The practice to identify and manage the communication needs of the project.

Communication requirements: The complete set of stakeholder communication requirements and needs.

Communication skills: The methods, procedures, and actions employed in the project to ensure that the information communicated is clear and properly understood.

Communication technology: Tools and techniques used to transfer information among project stakeholders (For example, emails and telephone).

Communication: Communication is the sending, receiving, processing, and interpretation of project information (For example, information can be transmitted verbally, non-verbally, actively, passively, formally, informally, consciously or unconsciously).

Communications planning: The process of planning project stakeholders' communication and information needs.

Compensable delay: A delay experienced by a subcontractor or an external resource for which a contractual compensation must be granted.

Compensation and evaluation: A performance measurement of a subcontractor or a resource and the financial reward as a result of that performance.

Compensatory time: A time off granted to a subcontractor or a resource by the project manager (For example, time off against overtime worked).

Competency: The critical skills needed to be able to do something successfully or efficiently (For example, the ability to lead a team is considered a critical skill for a project manager).

Competition: An activity or condition of striving to gain or win something by defeating or establishing superiority over others, an effort by providers of products or services to secure a project or parts of it independently

Competitive advantage: An advantage of one competitor over another. See also benchmarking.

Competitive benchmarking: A process of comparing a product or a service against several competitors using a set collection of metrics.

Competitive negotiation: A method of acquisition carried out by the buyer that allows bargaining and provides an opportunity for potential contractors to revise their proposals before contract award.

Completed activity: An activity that is finished with an actual finish date.

Completed: A project status describing project activities have been performed and accepted as complete by the customer or client.

Completion date: The date by which the project or a milestone is planned to complete.

Complexity estimate: A prediction model based on the numerical analysis of factors to forecast the level of complexity of the project.

Compliance: Adhering to specific standards, rules, or procedures established as necessary for operational effectiveness.

Compound Interest: A combination of interest earned on the principal in addition to a preceding period.

Compound risk: A risk resulting from several inter-related risks.

Compromising: A conflict management involves parties making concessions. See also Conflict management.

Computer software: A set of programs, tools, applications, and procedures designed to make use of the computer.

Concept phase: The first phase of the project lifecycle (For example, during this phase viability of the project is considered, and a favored solution identified).

Conceptual development: A process of selecting the best approach for achieving project objectives.

Conceptual estimate: See Order of magnitude estimate.

Conceptual solution: An initial project solution approach that may be required in the early stages of the project.

Concern: An expression of a problem or an issue where a lack of information about any aspect of the project may turn into a risk if neglected.

Conciliatory management style: A management style where the project manager is friendly and courteous. The intention is to unite all project parties involved to provide a compatible working team.

Concurrent engineering: An approach to project staffing, where project implementers are involved in the design phase. It is sometimes confused with fast-tracking.

Conditional diagramming method: A technique that represents a network and shows the repeating and conditional activities in the project (For example, this technique is valuable for helping you visualize and plan an activity's schedule).

Configuration identification: The activity which determines, defines, and documents the functional and physical characteristics and requirements of a system, including interoperability and interface requirements.

Configuration item: A part of a configuration that has a set function and is designated for configuration management.

Configuration management: The technical and managerial activities involved in the creation, maintenance, and controlled change of configuration throughout the life of the product.

Configuration status accounting: The activities concerned with recording and reporting of the current status and history of all changes to the configuration (For example, it provides a complete record of what happened to the configuration to date).

Configuration: The functional and physical characteristics of all project items, as defined in technical documents that should only be changed after authorization by the relevant manager (For example, documentation).

Configurations audit: an examination to make sure that all project deliverables conform with each other and to project specifications. It also ensures that relevant quality assurance procedures are implemented and that there is consistency throughout project documentation.

Configurations control: a system through which changes may be made to configuration items.

Conflict management: The ability to manage conflict creatively and effectively. A process of identifying and addressing differences and conflicts resulting in adopting opposing point views or positions (For example. Problem solving or confrontation, parties work toward a solution to the problem. Compromising, where parties make concessions to reach a standard solution. Forcing, when a project manager uses his or her authority to force a solution and Withdrawing is when one or both sides withdraw from conflict.

Conflict: A status of opposing sides resulting from incompatible expectations or views.

Confrontation: Also called problem-solving. See conflict management.

Consensus management style: A management style where the project team members work as a group to develop a solution and agree to support decisions made in the best interests of the project.

Consensus: Common agreement between the decision-makers that everyone can live with the solution (For example, all parties have to be convinced that the decision will achieve project objectives. If any party rejects the decision, then there is no consensus).

Consequential damages: Project damages that are proven to have occurred directly from one party's failure to meet a contractual obligation (For example, any injury to people or property resulting from any breach of warranty).

Constrained optimization methods: See Project selection methods.

Constraint: A condition or occurrence that regulates, limits, or restricts the project (For example, a target date).

Constructive cost model (COCOMO): An algorithmic software cost estimation model that uses a basic regression method, with parameters derived from historical project data as well as current project characteristics

Consultant: A subject matter expert, or solution specialist on a particular aspect of the project.

Consumable resource: A type of resource that will remain accessible until consumed (For example, a material).

Contingency allowance: See Contingency reserve.

Contingency plan: A formal document identifying project mitigation strategies to use should a risk event occur (For example, Examples a contingency reserve in the budget).

Contingency planning: The process of developing a plan that includes alternative strategies to minimize the adverse effects of a risk should it occur.

Contingency reserve: A designated amount set by the management or project manager to allow for future situations which may be partially planned (For example, sometimes called known unknowns as in a rework is required but the amount to carry out the rework is not known).

Contingency: A planned set of actions for minimizing the consequences of risk should the risk event occurs.

Contract Administration: The process of managing the relationship with the subcontractors and vendors.

Contract budget base: The estimated value of authorized but unpriced work on top of the agreed contract cost value.

Contract closeout: Conclusion and settlement of the contract, including the resolution of all outstanding items.

Contract dispute: A disagreement between contracting parties (For example, misinterpretation of technical requirements or terms and conditions).

Contract documents: All documents related to the contract include the contract itself and any support documents or set of documents that form the contract (For example, templates, appendices, financial schedules).

Contract financial control: The exercise of control over contract costs by the buyer or the contractor.

Contract guarantee: An assurance of performance of a contract by a contractor.

Contract negotiations: The process of discussing and compromising on contract terms to reach a final agreement between two or more parties (For example, clarification and mutual agreement on the structure and necessities of the contract before signing it).

Contract pricing: The practice of determining a specific pricing arrangement.

Contract quality requirements: The agreed quality requirements related to the product or service (For example, adherence to defined quality standard).

Contract target cost: The agreed cost for the defined contract and all contractual changes that have been agreed and approved (For example, the contract target cost equals the value of the budget at completion and the contingency reserve).

Contract target price: The negotiated estimated costs plus profit or fee.

Contract work breakdown structure (CWBS): Defines the scope of the effort and how the team will accumulate costs (For example, CWBS line up responsibility with accountability within the project team's organization and establishes the single numbering system that serves as the thread for the overall business information framework).

Contract: A contract is a mutually compulsory agreement that commits the seller to provide the specified product or services and commits the buyer to pay for it.

Contractor performance evaluation: A comprehensive review of the contractor's performance (For example, technical performance, cost performance).

Contractor qualification: A review of the contractor's past capabilities, experience, and past performance.

Contractor: A person or an organization that holds a contract for carrying out the works and/or the supply of products or services within a project. See also, Supplier.

Contractors conference: See bidders conference

Control Charts: Control charts are a graphic display of the results, over time, and against established control limits of a process (For example, control charts are utilized to ascertain if a process is in control or need of modification).

Control gate: A predetermined point in time during the project lifecycle for the project stakeholders to convene to determine project performance (for example, at the end of a milestone, or middle of the project) Also called Kill point, Stage exit or Phase end review.

Control limits: Control measurement bounties beyond which unacceptable performance is indicated (For example, upper control limit, and lower control limit).

Control: A process of comparing real project performance with scheduled project performance, assessing the difference for an appropriate corrective action.

Controlling processes: Actions agreed by the project team to ensure that project objectives are met (For example, monitoring and measuring progress and taking corrective action if needed).

Coordinated matrix: An organizational style where the project manager reports to the functional manager and does not have authority over team members from other departments.

Coordination: The act of ensuring that project work carried out by different organizations and in different places fits together effectively (For example, project technical aspects, time, and cost).

Copyright: Protection granted prohibiting others from reproducing author's and artist's work without formal permission.

Corporate culture: A reference to the behavior and belief that determine how an organization employees and management interact and handle professional transactions. Often, corporate culture is implied, not expressly defined, and develops organically over time from the cumulative traits of the people the company hires.

Correction: Elimination of a defect.

Corrective action: Changes made to bring an expected future performance of the project into line with the plan.

Cost account manager: A manager of a functional organization with the responsibility of cost performance, and the managing resources to accomplish such tasks.

Cost account: Defines the work is to be performed who will perform it and who is to pay for it.

Cost accounting practice: An established accounting method or technique used to measure cost, a framework used by the organization to estimate the cost of their products for analysis.

Cost analysis: The act of breaking down a cost summary into its elements for analysis and reporting.

Cost baseline: A time-phased budget used to assess and monitor cost performance on the project.

Cost-benefit analysis: An assessment of the relationship between the cost of undertaking a project, and the benefits expected to arise from the changed situation (For example, a comparison of the returns from alternative projects).

Cost breakdown structure: A hierarchical breakdown of a project into cost elements.

Cost budgeting: Assigning cost estimates to individual project parts.

Cost center: An identified location, person, activity, or project where costs may be established and recorded.

Cost code: A unique identifier for a specified activity that allows costs to be combined according to the foundations of a code structure.

Cost control point: The point in a program at which costs are entered and controlled (For example, the cost account or the work package).

Cost control system: A system of keeping costs within the budgets or standards based on work performed.

Cost Control: A management of changes to control project budget (For example, tools, techniques, and processes used to practice cost control, like variance analysis, cost and schedule reporting).

Cost curve: A graph that shows a horizontal time scale and cumulative cost vertical scale.

Cost-effective: Best value or performance for the smallest amount of cost.

Cost element: A unit of the cost related to a performance a task or to acquire an item (For example, a single value or a range of values).

Cost estimating relationship (CER): A mathematical relationship in which cost is expressed as a dependent variable of one or more independent cost-driving variables or as a function of one or more parameters (For example, parameters like performance).

Cost Estimating: Estimating the cost of the resources needed to complete project activities.

Cost forecasting: A process of predicting future trends and costs throughout the project.

Cost incurred: Costs identified as actually paid in the project (For example, direct labor, direct materials, and all allowable indirect costs).

Cost management plan: Part of the overall project management plan, a document that includes the processes, procedures, tools, and techniques required to describes how cost, budget, revenue, and expenditure, as well as cost variances, will be managed (For example, how to respond to a cost variance).

Cost management: A process of financial control of the project through evaluating, estimating, budgeting, monitoring, analyzing, forecasting, and reporting the cost information.

Cost objective: A cost object is used to describe something to which costs are quantified and assigned (For example, product lines, geographic territories, customers, departments).

Cost of capital: The opportunity cost of an investment. It is the rate of return that could be earned by investing the same money into a different project with equal risk.

Cost of quality: The cost sustained to ensure quality (For example, quality planning, quality control, quality assurance, and rework).

Cost overrun: The amount by which actual cost exceeds the planned cost on a project.

Cost performance index (CPI): A ratio of budgeted costs to actual costs to predict the amount of a possible cost overrun or underrun.

Cost performance measurement baseline: A budget costs and measurable goals in time and quantities for comparisons, analyses, and forecasts of future costs.

Cost performance report: A period cost report that reflects cost and schedule status information (or example, monthly or quarterly cost performance report).

Cost plan: A management budgeted amount, controls, and dates of incurring costs on the project.

Cost planning: The practice of planning the project cost estimating, evaluating, purchasing, and controlling procedures.

Cost-plus fixed fee (CPFF) Contract: A type of contract where the buyer reimburses the seller for allowable defined costs by the contract plus a fixed amount of profit (fee).

Cost-plus incentive fee (CPIF) Contract: A type of contract where the buyer reimburses the seller for allowable defined costs by the contract and a profit based on meeting defined performance criteria.

Cost-plus percentage of cost (CPPC) contract: Type of contract that provides reimbursement of the allowable defined cost of services performed plus an agreed percentage of the estimated cost as profit.

Cost reimbursement contracts: A type of contract based on payments to a contractor for allowable estimated costs, typically requiring only a 'best efforts' performance standard from the contractor.

Cost risk: Risk associated with failing to accomplish certain project activities within an estimated budget.

Cost-sharing contract: A contractual arrangement by which the contractor assumes some allowable defined contract cost.

Cost time resource sheet (CTR): A document that describes all significant elements in the work breakdown structure, resources required, the time frame of the work element, and a cost estimate.

Cost underrun: Amount by which actual project costs are below the planned cost.

Cost variance (CV): The difference between the estimated and actual costs of an activity.

Cost: Sum of cash or value of project activity or material, a value associated with resources spent to accomplish project objectives.

CPFF: See Cost plus fixed fee

CPI: See Cost performance Index

CPIF: See Cost plus incentive fee

CPM: See Critical path method

CPN: See Critical path network.

CPR: See Cost performance report.

Crashing: A project schedule technique used to compress the total project duration following the analysis of a number of alternatives to determine the maximum schedule duration compression for the minimum cost (For example, reducing schedule activity durations and increasing the assignment of resources on scheduled activities.

Criteria: A set of guidelines and procedures to be followed for project development, design, or implementation.

Critical activity: A critical activity with zero or negative float (For example, an activity that has no allowance for work slippage that must be finished on time or the whole project will fall behind schedule).

Critical issue: A project problem that cannot be ignored and must be addressed before proceeding to the next phase.

Critical item: A project item or component that could jeopardize the completion of project objectives if deemed not available when required (For example, items that could harm cost, schedule, quality).

Critical path analysis: A procedure for calculating the critical path and floats in a network.

Critical path method (CPM): A network analysis method to predict project duration by analyzing the sequence of activities on the path with the least amount of scheduling flexibility or float. Early dates are calculated utilizing a forward pass using a specified start date. Late dates are calculated employing a backward pass starting from a specified completion date.

Critical path network (CPN): A sequence of project network activities that sum up to the longest overall project duration, irrespective of the longest duration, including a float or not. This provides for the shortest time possible to complete the project.

Critical path: The path in a project network that has the longest duration. This represents a series of activities that determines the earliest completion of the project.

Critical performance indicator: A critical factor against which aspects of project performance may be assessed.

Critical risk: A risk that can jeopardize the achievement of a project's objectives.

Critical success factor: A factor considered to be most beneficial to the achievement of a successful project.

Criticality index: A tool used in risk analysis that represents the percentage of ratio resulting from an activity presence on the critical path (For example, the criticality index of an activity may be articulated as a ratio between 0 and 1 but is more often expressed as a percentage).

Cumulative cost curve: A graphic display used to show planned and actual expenditures to monitor cost variances.

Current costs: A present market value of a product or service.

Current finish date: The present estimate of the time when an activity will be completed.

Current liabilities: The sum of debts incurred by a project or an organization which are expected to be paid

Current start date: The present estimate time when an activity will begin.

Customer acceptance: A formally documented signoff by the customer that project deliverables satisfy requirements.

Customer: A person or an organization representative who defines the needs of the project, justifies and pays for the project (For example, client, owner).

Cut-off date: The ending date of a reporting period.

Cutoff point: The minimum acceptable rate of return an organization is willing to earn on its investments.

Cutover: The process of shifting from one system to another (For example, the two systems are not functioning in parallel).

CV: Cost Variance

CWBS: See Contract work breakdown structure.

D

Dangle: An activity with either no predecessors or no successors. If neither exists, it is referred to as an isolated activity.

Data date (DD): A point in time that separates actual data from future data. Also called as of date.

Data refinement: An exercise of redefining data that was developed during project planning (For example, refining data for a project milestone and resources).

Data: Documented project information, facts, and statistics collected together for reference or analysis, regardless of the form or the media on where it is recorded.

DCF: See Discounted cash flow.

DD: Data date

De facto authority: A term used to describe a person or a group that has control but may not have the legal authority (For example, a person exercising influence regardless of formal power).

Decentralized contracting: Contract management arrangement in which the project manager typically has control over the contracting process for his or her project.

Decision event: A state in the progress of a project when a decision is required before the start of any following activity (For example, the decision determines which of several alternative paths is to be followed).

Decision making: The act of analyzing a problem to identify working solutions and then making a choice.

Decision support system: A computer software used to aid in decision making (For example, Simulation software, and mathematical routines).

Decision theory: A technique used in risk quantification, involves calculating consequences of indeterminate decisions to assist in decision making (For example, calculating the expected value of risk, and pointing to the best possible course of action).

Decision tree: A flowchart structure in which each internal node represents a "test" on an attribute (For example, if a coin toss comes up head or tail), each branch shows the outcome of the test, and each leaf node shows a class label.

Decision: The removal of uncertainty concerning a course of action.

Decomposition: Subdividing project deliverables into smaller, more manageable components into components defined in sufficient detail.

Decrement: Cost or price reduction.

Default: Failure to perform a project contractual obligation.

Defect: Nonconformance, or defects in units of any given quantity of product.

Definitive estimate: An accurate, detailed estimate within -5 to +10 percent that is prepared from defined specifications (For example, the estimate used to answer RFPs or bid proposal).

Deflection: A transfer of project risk to a third party (For example, insurance or warranty).

Delaying resource: In project resource scheduling, insufficient availability of resources may warrant that the completion of an activity be delayed beyond the date on which it was planned be completed (For example, the delaying resource is the first resource on an activity that causes the activity to be delayed).

Delegation: The practice of effectively getting project team members to perform work (For example, responsibility is distributed from project manager to team members).

Deliberate decision event: A decision event made as a result of the outcomes of the preceding activities and possibly other information.

Deliverable: Any measurable, tangible, and verifiable item that must be produced to complete a process, phase, milestone, or project.

Delphi technique: A process where a consent view is reached by consultation with experts (For example, the techniques are used as a project estimating technique or project risk identification).

Democratic management style: A management style that largely depends on stakeholder's participation (For example, the project manager and project team can participate in making project decisions).

Demonstration: An exercise of verifying compliance with project specifications (For example, a project team witnessing how a procedure works).

Denied: Project status describing proposed project work that will not be considered.

Dependability technique: A method of identifying improvements by analyzing a system's behavior to increased reliability.

Dependency arrow: A link arrow on an activity on a network to represent the interrelationships of activities in a project.

Dependency: A relationship activities or project tasks. A dependency may be logical or resource-based. See Logical relationship.

Dependent activities: Activities that are related in a way that the beginning or end of one activity is contingent on the beginning or end of another.

Depreciation: A reduction in the value of an asset over time, due in particular to wear and tear.

Design/build: A construction method involves the combination of project architectural, engineering, and construction services required for a project into a single agreement.

Design and development: System design is the process of defining the elements, interfaces, modules, and data for a system to satisfy stated requirements. System development is the process of generating or altering systems, along with the processes, models, and methodologies used to develop them.

Design authority: The person or organization with overall design responsibility for the products of the project

Design of experiments: The design of a task that aims at describing the variation of information under certain conditions to reflect the variation (For example, an experiment in project quality management to help identify the variables that have the most influence on the outcome of a process).

Design review: A formal, documented examination of a system design to evaluate its capability to meet requirements.

Design specification: A detailed document providing a list of points regarding a product or process (For example, the design specification could include required dimensions, environmental factors, ergonomic factors, aesthetic factors, maintenance that will be needed).

Design to cost: A cost management techniques, describes a systematic approach to controlling the costs of product development and manufacturing by designing cost into the product at the earliest concept decisions and are difficult to remove later (For example target cost goals are established during the development of a system by controlling of costs, acquisition, operations, and support).

Design: A specification or drawing description of a product or service to provide required information on how to build a product or perform a service.

Detailed design: An output of system design; a detailed, component-level technical or engineering description of a system (For example, detailed information on specific applications in a project).

Detailed Implementation Plan: The updated schedules, plans, and targets that have sufficient detail to allow the effective project implementation (For example, a detailed implementation plan is completed following a proposal approval and before implementation begins).

Detailed schedule: A comprehensive schedule used to describe the day-to-day activities of the project.

Deterministic estimate: a predetermined estimate with a defined result.

Deterministic network: A display diagram that includes paths that must be followed with fixed durations (For example, the deterministic network is used to differentiate between traditional networking from probabilistic networking).

Development goals: The fundamental basis for which a project is undertaken

Development methodology: A set of integrated processes and procedures organized into a series of phases creating the development cycle of a product or a service.

Development phase: A phase in the project lifecycle, where project planning and design typically occur.

Deviation permit: A formal authorization to depart from a requirement or specification for a specified item and time. See also production permit.

Dialogue: A dialog where participants share their thoughts and enhance an understanding of the subject and, possibly, reach consensus.

Direct costs: Costs associated explicitly with an activity or group of activities.

Direct labor: Labor that is identified with a specific cost objective (For example, direct engineering labor includes design, quality assurance, and testing).

Direct materials: Parts, equipment, and other items purchased to manufacture or assemble completed products.

Direct overhead: The sum of overhead costs that can directly be attributable to a project (For example, office space, insurance, and utilities).

Direct project costs: Costs directly associated with a project (For example, including all personnel, goods, or services).

Directive: A formal project communication that initiates or prescribes and action.

Disallowance: A rejection of a project cost as an allowable cost.

Discipline: An Area of technical expertise or project specialty.

Discount rate: The interest rate that is used in calculating the present value of future cash flow. See also Discounted cash flow.

Discounted cash flow (DCF): A concept of relating future cash inflows and outflows over the life of a project, a financial technique for calculating the present value of future expected expenses and revenues using net present value.

Discrete milestone: A milestone that has a scheduled occurrence in time, but no other resource.

Discrete work package: A short term project activity with a definite start and endpoint that may be used to measure performance or earned value.

Discretionary dependency: A dependency defined by preference compared to necessity. Also referred to as Preferred logic or Preferential logic, or Soft logic (For example, based on a "have to," but on a "should").

Dispute: A none settled disagreement by mutual consensus that could be decided by litigation or arbitration.

DU: Duration

Dummy activity in activity on arrow network: An activity of zero duration with no actual work to be done but may be needed for reasons of logic (For example, there are three uses for a dummy activity in 'activity on arrow network': logic, time delay).

Duration (DU): Number of work periods required to complete an activity or a project milestone (For example, durations are expressed in hours, workdays, or workweeks).

Duration compression: Shortening the project schedule without reducing the project scope. Also called Schedule compression. See also Crashing and Fast-tracking.

E

EAC: Estimate at completion

Earliest feasible date: The earliest date when the activity could be scheduled to begin based on the scheduled dates of all its predecessors.

Early dates: The early date on which an activity can start and finish.

Early finish date: The earliest estimated date on which an activity can end. It is based on the activity's Early Start, which depends on the finish of predecessor activities and the activity duration.

Early Start Date (ES): In a critical path method, the earliest possible time when the unfinished part of an activity (or the project) can start

Early warning system: A project control monitoring and reporting system used to alert the project manager if trouble is about to arise.

Earned hours: The time in hours credited from the completion of a given task or a milestone.

Earned value (EV): An approach that involves monitoring the project plan, real work, and work accomplished value to see if a project is on track (For example, EV shows the amount and time that should have been spent, considering the amount of work done so far).

Earned value analysis: Analysis of project progress that involves the actual money budgeted and spent compared to the value of the work achieved (For example, actual money spent or budgeted per work period in days, weeks, or months).

Earned value cost control: The quantification of the progress of a project in financial terms to provide a realistic measurement of comparing the actual cost to date.

Earned value management: A project control process based on a structured approach to planning, cost collection and performance measurement (For example, EVM enables the integration of project time, scope, and cost objectives, and the establishment of a baseline plan for performance measurement).

E-commerce: Conducting business transactions between businesses or between businesses and consumers usually over the internet (For example, buying products online, paying bills online, or carrying out banking transactions online).

Economic evaluation: A process to measure the value of a project concerning specific standards or benchmarks.

Economic lot size: Quantity of material or units of a manufactured product that can be produced or purchased within the lowest unit cost range (For example, reconciling the decreasing unit cost of more substantial quantities with the associated increasing unit cost of handling, storage, insurance).

Economy of scale: A proportional saving in costs by an increased level of production.

EF: Early Finish date

Effective team: A project team who work together in a constructive way to accomplish project objectives.

Efficiency factor: The relationship between the allowance time and the time taken in percentage (For example, performance rating and remuneration calculation exercises).

Effort driven activity: An activity whose duration is governed by resource usage and availability (For example, the resource requiring the longest time to complete the work on the activity will determine its duration).

Effort remaining: An estimate of effort remaining to complete an activity, a milestone, or a project.

Effort: The number of work units necessary to complete the project (For example, man-hours man-days or man-weeks).

Egoless team structure: A team structure where there is no apparent leader, decisions are made through consensus, and project work tends to reflect the input of all team members (For example, when the team is small with clear vital objectives).

Eighty-hour rule: A method of breaking down each project activity into work packages that require a max of 80 hours of effort to complete.

Elaborated: Planned and developed thoroughly (For example, activities planned details).

Elapsed time: The total number of calendar days that is needed to complete an activity (For example, workdays not counting weekends or holidays).

End activity: An activity with no rational successors.

End event: An event with a proceeding activity, but no following activities (For example, there could be more than one end event.

End product: The deliverable from project work.

End-user: The person or organization for whom the project's product or service is intended.

Endorsement: A formal approval of understanding and acceptance of the project objectives and deliverables.

Engineering buildup estimate: See Bottom-up estimate.

Engineering change order: An order to include project improvements that have been designed following the initial design.

Engineering estimate: See Definitive estimate.

Engineering review board (ERB): A technical oversight committee of functional organizations staff involved in aiding the project manager.

Enhancement: Changes to the scope of an existing project to provides additional functionality or features.

Enterprise model: A description of a business aspect or organization (For example, vision, mission, objectives, processes, and business activities).

Enterprise project management office (EPMO): A centralized business function that operates at a strategic level with the enterprise executives and provides enterprise support on governance, project portfolio management best practices, mentoring, tools, and standardized processes. See also Program management office and Project Management Office.

Enterprise: Company, organization, or a group of companies or organizations.

Environmental factoring: Data relating to an external factor to modify the value of parameters concerned (For example, Weather).

Equipment procurement: Acquisition of equipment or material for the project.

Equitable adjustment: An adjustment, usually used in government contracts according to a change's clause, compensation expenses incurred due to actions of the project owner.

Equivalent activity: An activity that is equivalent to any combination of series and parallel activities.

ES: Early start date

Escalation: A surge in the intensity or significance of a project issue (For example, escalating a project risk to the project steering committee due to its seriousness of serious impact).

E-Signature: Data in electronic form, which is logically associated with other data in an electronic form used by the signatory to sign.

Estimate at completion (EAC): A value expressed in money or units, to represent the expected final costs of work when completed.

Estimate to complete (ETC): The value expressed in money or units developed to represent the cost of the work required to complete an activity or a project.

Estimate: An assessment of the required duration, effort, or cost to complete a task or project (For example, estimates are not actual, they should always be expressed with some indication of the degree of accuracy).

Estimated cost: Anticipated cost of performing an activity, a milestone, or a project.

Estimating guidelines: The procedures set by an organization for estimating project work effort, cost, or schedule (For example, formulas and criteria for assessing the factors affecting the cost estimate).

Estimating: An approximation using a range of tools and techniques to produce estimates of project costs and targets that are refined throughout the project lifecycle.

ETC: Estimate to Complete

Ethical management style: A management style in which the project manager is honest, sincere, and able to motivate the project team towards the best solution.

EV: Earned value

Evaluate: Appraise against a set value.

Evaluation criteria: The basis used to measure and score proposals submitted by prospective contractors (For example, criteria may be objective or subjective).

Evaluation: A periodic assessment of a project's efficiency, effectiveness, and impact on a defined population.

Event chart: See Milestone chart.

Event on Node: A network diagramming technique in which events are represented by boxes or nodes connected by arrows to show the sequence of events.

Event: A defined point is the beginning or end of an activity.

Evidence-based reporting: An approach to report writing where statements made about the progress of the project are supported with verifiable information.

Examination: An inspection of products and services to determine conformance to specified requirements.

Exception report: A type of reports designed to draw attention to instances where planned and actual results are different (For example, an exception report is usually caused when actual values may be better or worse than the plan).

Exceptions: Occurrences causing deviation from a plan (For example, issues, change requests, and risks or items where the cost variance and schedule variance exceed predefined thresholds).

Exclusive OR relationship: Logical relationship showing that only one of the possible activities can be undertaken.

Executed contract: A signed contract where the terms and conditions have been fulfilled by the contractual parties.

Executing processes: Activities associated project plan implementation.

Executing: The process of coordinating all project resources in the performance of the project or the actual performance of the project.

Execution phase: The phase of a project in which project activities towards the direct achievement of the project's deliverables and objectives occur. Sometimes called the Implementation phase.

Expectancy theory: A theory of motivation involves people will tend to be highly motivated and productive if they believe that their efforts will likely lead to successful results, and they will be rewarded for their achievement.

Expectations: Predicted changes in performance as a result of project implementation (For example, productivity, and operations).

Expected monetary value: The result of an event's probability of occurrence and the gain or loss that will result (For example, if there is a 20 percent probability that an event will occur, and the event will result in a $1000 loss, the expected monetary value of the event is (.2 x $1000).

Expected value: The result of multiplying the probability of a risk occurrence with its estimated monetary impact.

Expenditure: A financial charge against available funds (For example, a voucher, a claim, or other documents).

Expert authority: Influence that is derived from one's knowledge or expertise, rather than from some outside source. Also called technical authority.

Expert judgment: The result of a requested recommendation or advice by a person or an organization recognized as having specialized knowledge in a specific area.

Expert power: See Expert authority.

Expert system: A computer-based system that bases decisions on the knowledge of recognized experts in a certain (For example, rules, parameters, and input are used to derive decisions for specific project problems).

Exposure: The likely loss or consequence of a risk (For example, the combined probability and impact of a risk).

External audit: Audit performed by an entity that is external to the project team.

External constraint: A constraint originating outside the project network.

External dependency: The relationship between project activities and non-project activities. Such dependency involves things that are beyond the control of the project team but should be reflected in the project schedule.

External feedback: Information about the performance of the project team from individuals or entities outside the project.

External risk: A risk beyond the control or influence of the project team.

F

FAC: See Forecast at completion.

Facilitating management style: A management style where the project manager makes himself or herself available to answer questions and provide guidance when needed.

Facilitator: A person assigned to help the project teamwork more effectively (For example, a person facilitates an exercise for project risk identification).

Fair and reasonable price: A price considered to be fair to both parties based on agreed conditions, quality, and performance.

Fair market price: A price at which real sales have been made in a particular market at the time of purchase.

Fallback plan: A plan for an alternative set of actions that can be implemented to overcome the consequences of a risk (For example, carrying out any agreed activities that may be required to render the plan active).

Fast decision process: A process in which an authorized individual or team makes decisions quickly.

Fast-tracking: Av exercise of reducing the duration of a project by overlapping activities or phases originally planned to be performed sequentially (For example, reducing the number of sequential relationships and replacing them with parallel relationships to achieve shorter overall durations but often with increased risk).

Feasibility estimate: See order of magnitude estimate.

Feasibility phase: See Concept phase.

Feasibility phase: The project phase that demonstrates that the requirement can be achieved, this phase identifies and evaluates the options to determine the solution.

Feasibility Study: Analysis to examine the viability of taking on a project.

Fee for professional services: Amount paid for professional services for work performed satisfactorily.

Fee: Amount paid to the contractor beyond stated costs in a cost-plus fees contract.

Feedback: Information extracted from a project activity or situation and used to modify immediate or future processes or situations (For example, feedback from a milestone acceptance testing used to modify future testing).

FF: See finish to finish.

Field cost: Cost associated with establishing, operating, and maintaining the project site.

Fifty-fifty progress reporting: A statistical approximation approach of the budgeted cost of work performed. The 50-50 progress reporting is a method to determine what percentage of the task has been completed, where half the effort is assumed to be completed, and half the budgeted cost of work scheduled associated with the task is entered into the project account book. The other half is entered once the work is complete.

Final completion: A formal approval that the entire project is completed and in accordance with the agreed requirements and scope (For example, approval that the project is complete and can be moved to operation and warranty stage).

Final report: A post-implementation report that formally closes the project to be handed over the project deliverables for operational use.

Financial closeout: A formal close of the project accounting systems signaling that no further charges should be made against the project.

Financial control: See cost Control.

Financial statement: A formal record documenting the financial status of the project or organization that includes assets and liabilities.

Financing: Raising required funds for the project (For example, project owner's funds, stocks, and bonds).

Finish date: The actual or estimated date associated with the completion of an activity.

Finish Float: Finish float is the amount of excess time activity has at its finish before a successor activity must start.

Finish to finish (FF): Relationship in a precedence diagramming method network in which one activity must end before the successor activity can end. See also logical relationship. See logical relationship.

Finish to finish lag: The finish to finish lag is the minimum amount of time that must pass between the finish of one activity and the finish of its successor(s).

Finish to start (FS): Relationship in a precedence diagraming method network where one activity must end before the successor activity. See Logical relationship.

Finish to start lag: The minimum amount of time that must pass between the finish of one activity and the start of its successor(s).

Finishing activity: The last activity that must be performed before a project can be considered complete (For example, this activity is not a predecessor to any other activity, and it has no successors).

Firm offer: A formal offer from a seller to a buyer that is valid for a set period of time.

Fiscal Year: A defined 12-month period (For example, January 1 to December 31).

Fishbone diagram: See Ishikawa diagram.

Fixed asset: Property or equipment used for the construction of products and services (For example, real-estate, and machines).

Fixed cost: Project expenses that are not dependent on the level of products or services produced by the project. They tend to be time-related (For example, interest or rents being paid per month).

Fixed date: A calendar date that is associated with a project plan which cannot be moved or changed during the schedule.

Fixed duration scheduling: A scheduling method where the duration remains the same regardless of the number of resources assigned to the project task.

Fixed finish: See imposed finish.

Fixed price (FP) contract: A type of contract where the buyer pays the seller a set and agreed to amount regardless of the seller's costs.

Fixed price plus incentive fee (FPIF) contract: A type of contract where the buyer pays the seller a set and agreed to amount regardless of the seller's costs plus an incentive-based on agreed performance criteria.

Fixed start: See imposed start.

Float: It is the difference between the task's early and late start dates; the amount of time available for a task to slip before it results in a delay of the project end date.

Floating task: A task that can be performed anytime without affecting the project duration or critical path.

Flow diagram: A graphic representation of the logical sequence and flow of the project tasks and activities (For example, the diagram shows the logical flow of tasks with no regard to their duration).

Flowchart: A type of diagram that represents a workflow or process, can also be defined as a diagrammatic representation of an algorithm (Fr example, a step-by-step approach to solving a project problem showing the steps as boxes of various kinds, and their order by connecting the boxes with arrows).

FOB: See free on board.

Follower: A project task that logically succeeds another task.

Follow-up audit: An audit performed to determine that the recommendations resulting from a previous audit were implemented effectively.

Force field analysis: Quality technique that can be used for identifying, discussing, and documenting the factors that support or oppose a change initiative.

Forcing: See conflict management.

Forecast at completion (FAC): Scheduled cost for a task. See estimate at completion.

Forecast final cost: See estimate at completion.

Forecast: Prediction of future conditions and events based on information and knowledge available at the time of the estimate.

Forecasting: Predicting future conditions and events done during the planning process.

Foreign currency exchange: Financial instruments such as paper currency, notes, checks, and bills of exchange used in making project payments.

Formal acceptance: The formal process of accepting delivery of a deliverable.

Formal authority: A type of influence based on an individual's position in the organization or conferred by the organization. Also called Legitimate authority.

Formative quality evaluation: A method for evaluating the quality practices that focus on process in a project while the project activities are in progress.

Forming: Typically, a festive period as the team is orientated with the project. A general sense of uncertainty is expected as people try to come to terms with what the project is about and what their roles will be. Questions like "do I want to be part of this group?" are common. In this stage, team members would naturally defer to the project leader or dominant member for guidance; the team is looking for someone to take charge and give it the information it needs, a decisive leadership if you like

Formula estimate: A formula-based estimate used to calculate the total estimated effort for a task or work breakdown element (For example, the variables in the formula are count, Low, and High provided by one or more estimating factors).

Forward pass: The calculation of the early start and early finish dates for uncompleted portions of all activities. See also Network analysis and Backward pass.

Forward scheduling: Planning the tasks from the date resources become available to determine the due task date.

FPIF: Fixed price plus incentive contract

Fragment: See Subnet.

Free float (FF): The amount of time an activity can be delayed without delaying the early start of any immediately following activities. See also Float.

Free on board (FOB): A pricing arrangement where the seller agrees to bear the shipping cost to a specific location.

FS: Finish-to-start

Full audit: Audit of all aspects of the project.

Function point analysis: A method of Functional Size Measurement to assesses the functionality delivered to its users, based on the functional requirements.

Function quality integration: Quality practices that are integrated seamlessly with standard business processes to ensure that quality plans and programs are integrated, consistent for the project team to achieve standard product quality.

Functional department: A department within an organization assigned to perform a specific function (For example, engineering, manufacturing, marketing).

Functional Manager: A manager responsible for the activities of a particular functional department. (For example, engineering, manufacturing, marketing). See also Line manager

Functional matrix: An organization structure where the project has a team leader in each functional department, and the products are passed from one team to the next.

Functional organization expert: An individual who is process or knowledge experts to provide validation of certain aspects of the project.

Functional organization: A management structure where specific functions an organization are grouped into specialist departments providing dedicated services (For example, marketing human resources, finance).

Functional requirements: Features and functions of a product or a project deliverable described in language that is understandable to the end-user.

Functional specification: A formal document describing the required functions of a system.

Funding profile: An estimate of funding requirements over time.

Funding: An organizational process confirming approval and allocation of financial resources for a project to proceed.

G

G&A: See General and administrative expenses.

Gainsharing: A management mechanism to encourage employees or team members to become more productive, motivating the team to boost their performance through profit participation and involvement.

Gantt chart: A bar chart showing a schedule of project activities and milestones. Activities are typically listed with other tabular information along the left side of the chart with a timeline across the top or bottom.

Gap analysis: Involves the comparison of actual performance with potential or desired performance, or an investigation of the difference between the present state and the desired state of a deliverable or a system.

General and administrative (G&A) expense: Expenses allocated to an organizational unit for the general management and administration of the organization.

General management: A reference to the different aspects of managing an organization.

General provision: A scheme where employees or team members receive a bonus (For example, a percentage of salary based on specific performance criteria).

General requirements: Project requirements that are non-technical or specific to any product or service (For example, payment terms).

General sequencing: An outline of the order in which activities are performed.

GERT: Graphical evaluation and review technique.

Go no go: A Determination to in the project lifecycle to proceed with or abandon a plan or project the measure allows a project manager to decide whether to continue, change, or end an activity or project (For example, if a project is facing a grave financial or legal risk).

Goal statement: A high-level statement of the project's objective or purpose,

Goal: The objective that the project is intended to accomplish.

Gold plating: Intentionally adding extra features or functions to the products which were not included in the scope statement at no additional cost to the customer.

Governance model: The set of agreed processes, roles, and responsibilities for governing the progress and direction of a project.

Governance of project management: Involves the areas of organizational governance that are specifically related to project activities.

Governance: The planning, influencing guidance of the policies and activities of the project.

Gozinto chart: A vertical tree diagram that displays hierarchical levels of detail of a complete product assembly (For example, this project planning tool is of value for a bill of materials (BOM).

Grade: A category or rank to distinguish items with the same functional use but different requirements for quality (For example, different types of wood may need to withstand different amounts of force).

Grant: Funds provided to an organization (For example, funds to a non-profit organization of local government to carry out a project).

Grapevine: Information passed, heard or communicated informally or unofficially, usually learned by word of mouth (For example, project gossip that may be true but usually incomplete).

Graph: A graphical display of project activity relationship (For example, histograms and control charts).

Graphical evaluation and review technique (GERT): A network analysis technique that allows for conditional and probabilistic treatment of logical relationships (For example, to elect that some activities may not be performed). See also Conditional diagramming method.

Grassroots estimate.: See Bottom-up estimate.

Gross profit: The difference between earned revenue and direct cost of products or services.

Group communication: Communication shared during project meetings, consultations, and presentations to convey information to the project stakeholders.

Group dynamics: The behaviors and psychological processes occurring within the project team (For example, positive dynamics or negative dynamics depending on the personality characteristic of the people in the group.

Groupware: A description of any software application that runs on a network that allows groups of people to work collectively and collaboratively.

Guideline: A document that recommends techniques and procedures to be used to accomplish a task.

H

Hammock: An aggregate or summary activity (For example, a group of related activities is shown as one and reported at a summary level). See also subproject and subnet.

Handover: The formal process of transferring responsibility and ownership of the products of a project to the end-user.

Hanger: An unintended break in a network path caused by missing activities or missing logical relationships.

Hard copy: A printed copy of the information.

Hard logic: See Mandatory dependency.

Heuristic: Enabling a person to discover or learn something for themselves, a problem-solving technique that results in an acceptable solution.

Hidden agenda: Unknown objectives of a person or group to others during the course of a project to work against project objectives and to undermine the stated project objectives.

Hierarchical coding structure: A coding system represented as a multi-level tree structure in which every code except those at the top of the tree has a parent code.

Hierarchical management: A traditional line management structure in which functional areas are created and staffed with human resources.

Hierarchy of networks: Range of networks showing the relationships between those networks at different levels of detail, from summary down to working levels.

High-performance teams: A group of people with specific roles and complementary talents and skills, aligned with and committed to a common objective (For example, teams that show high levels of collaboration and innovation, produce superior results).

Histogram: a graphic display of planned and or actual resource usage over a period of time.

Historical records: Previous project documentation, usually collected from lessons learned (For example, such documentation is used to help predict trends, feasibility, or risk identification and mitigation on future similar projects.

Historical cost: Actual cost incurred in performing the project or part of it.

Historical estimating: An estimating technique of work effort and costs using documented data from past similar projects.

Holiday: A none working day as recognized on the project calendar.

Holistic: A view of the entire effort rather than each activity or task individually.

Host organization: The organization responsible for providing administrative and logistical support for the project.

Human resource gantt chart: A graphically illustrates how project resources are allocated to tasks and distributed throughout the life of a project. A variation of the horizontal bar Gantt chart.

Human resource loading chart: Vertical bar chart used to show personnel resource consumption by time period.

Human resource management: The understanding and application of the policy and procedures that directly affect the people working within the project.

Hypercritical activities: Activities on the critical path with negative float.

I

Idle time: A time interval during which the project team does not perform project activities or tasks.

IFB: Invitation for Bid

Impact analysis: Assessing the advantages of pursuing a particular course of action.

Impact: Positive or negative long-term consequence assessment of an occurring risk.

Implementation phase: The project phase that develops the agreed solution into a completed deliverable.

Implied warranty: A contract term for certain assurances that are presumed to be made in the sale of goods due to the circumstances of the sale (For example, goods or products must be reasonably fit for the ordinary purposes(s) for which they are used).

Imposed finish: A finished date imposed on a project activity by external constraints.

Imposed start: A start date imposed on a project activity by external constraints.

In-house: Work performed by resources within the performing organization.

In progress activity: An activity that has started but not yet completed.

In progress: A project activity that has been started but not yet completed.

In-service date: The date by which the product or service is ready and available for use.

Inaccuracy allowance: A provision of time or money for possible schedule or cost estimates inaccuracies.

Incentive contract: See Fixed price plus incentive contract.

Inclusive OR relationship: A logical relationship indicating that at least one but not necessarily all of the activities have to be performed.

Incremental delivery: A project lifecycle strategy used to reduce the risk of project failure by dividing projects into more manageable pieces (For example, the resulting sub-projects may deliver parts of the full product or solution).

Incurred costs: Sum of actual and dedicated costs at a specified time.

Indemnification: A contractual obligation of one party to compensate for the loss incurred to the other party due to the acts of the Indemnitor or any other party.

Independent cost estimate: Project cost estimates conducted and provided by external individuals or entities.

Independent verification and validation: A verification and validation process provided by an external individual or entity other than the one that designed or implemented the solution

Indirect cost: A set of costs allocated to a project but are not directly associated with an activity or group of activities (For example, overhead cost, administration cost or a burden rate of $0.25 which means you spend $0.25 on indirect labor costs for every dollar of gross wages you pay).

Indirect prejudice: The cost that results from a product failure to meet specification or project performance criteria or failure to meet contractual commitments.

Inflation: A sustained surge in the general price level of goods and services over a period of time (For example, taking into consideration factors such as cost of living and interest rates).

Influencing the organization: The skill an individual possesses to make things happen or get things done

(For example, familiarity with formal and informal structures of all the organizations and politics).

Information Distribution: Making needed information available to project stakeholders in a timely manner.

Information management and reporting: The collection, dissemination, distribution, and archiving of project information. Reporting is presenting project information in an appropriate format to stakeholders.

Information overload: Exposure to quantity, type, and complexity of information to a point where the ability to comprehend and use such information is reduced.

Information requirement: Information needed to perform an activity or a project.

Information system: A design of complex formal and organizational systems designed to collect, process, store, and distribute information for use in decision-making activities.

Initial project plan: A high-level plan used to document the conceptual approach to a project.

Initiation: Committing the organization to start a project or phase, a formal recognition that a project exists or that an existing project should continue to next phase.

Input limits: Restrictions imposed on resources needed to perform project activities and tasks.

Input priorities: Activities that require schedule urgencies within forced constraints.

Input restraint: Required external constraint (For example, dates and float allocation).

Input: Information and items required for an activity to start (For example, the financial, human, material, and information resources used for the development).

Inspection by attributes: Examining an item, or characteristics of an item, and classifying it as "conforming" or "non-conforming." Attributes control is at the limits, variables control within limits. Concerning the data that is generated by each concept, attributes data is discreet, whereas variables data is continuous.

Inspection by variables: A quality control inspection method in which the sampled articles are evaluated on the basis of quantitative criteria. See Inspection by attributes

Inspection cycle: Series of inspections to ensure that products are kept in an operational condition.

Inspection in process: A preventive examination during product manufacturing to inspect the features and characteristics that cannot be examined during the final inspection.

Inspection level: An inspection of the entire sample size for quality issues related to performance, function, and visual appearance.

Inspection record: Information related to the results of inspection action.

Inspection system requirement: Specific inspection requirements to carry out and maintain conformance to particular specifications.

Inspection: Examination or measurement of an activity or project to validate conformance to a specific requirement.

Insurable risk: A project risk that can be covered by an insurance policy. Also called Pure risk.

Insurance: A premium amount paid by the project to cover the cost of a risk impact. See also Impact.

Integrated cost/schedule reporting: See Earned value.

Integrated logistics support (ILS): All the considerations necessary to ensure effective and economical support of a system over its lifecycle (For example, planning, supply maintenance, and support).

Integrated project progress report: A comprehensive report of the project progress that involves cost, revenue, performance, resources, schedule, risk, quality, among others.

Integration management plan: Part of the overall project management plan, a document that includes the processes, procedures, tools, and techniques for making sure that all project elements are appropriately coordinated.

Integration management: A process of bringing people, activities, and all other project elements together to perform effectively.

Integration planning: The practice of incorporating people, activities, and all other project elements together to perform effectively.

Intellectual property: A category of property that includes intangible creations of the human intellect (For example, Concept, idea, thought, drawings, or process).

Intention to bid: A written communication by potential contractors indicating their willingness to perform the specified work.

Interdependencies: The relationships among organizational functions where one function, task, or activity is dependent on others.

Interest: Money paid regularly at a particular rate for the use of money or for delaying the repayment of a debt.

Interface activity: An activity connecting two nodes from two different subnets to represent logical interdependence.

Interface management: Essentially, the project manager's job: planning, coordinating, and controlling the work of others at project interfaces (For example, management of communication, coordination, and between project stakeholders).

Interfaces: Boundaries between functions.

Internal audit: An audit performed by the project team or an internal function within the organization.

Internal control: An internal process of monitoring and dealing with deviations from the project objectives and plan.

Internal project sources: Archived and historical information within the organization on similar projects.

Internal rate of return (IRR): A discount rate at which the net present value of future cash flow is zero (For example, IRR is a particular case of the discounted cash flow).

Internal risk: A risk that is within the control and influence of the project team. See also External risk.

International standards organization (ISO): An international standard-setting body composed of representatives from various national standards organizations.

Internet portal: See Portal.

Interpersonal interfaces: A reference to the formal and informal reporting relationships among project stakeholders.

Intimidating management style: A management approach in which the project manager frequently admonishes team members (For example, a demanding manager maintains his/her image at the risk of depressing team morale).

Inventory closeout: Payment and credit of inventory purchased by project funds.

Inverted matrix: A project-oriented organization structure that employs permanent specialists to support projects.

Invitation for bid (IFB): Equivalent to request for proposal. An invitation to prepare a detailed proposal based on a specific request or a tender.

Invoice: A contractor's itemized statement addressed to the purchaser of products shipped or services performed (For example, a detailed and itemized bill with quantity and prices).

IRR: See Internal rate of return.

Ishikawa diagram: A diagram used to illustrate how causes create a specific effect. Also called a cause-and-effect diagram or fishbone diagram.

ISO: See International standards organization.

Issue management: The process by which concerns that threaten the project objectives are managed.
Issue: An immediate problem that requires a resolution (For example, a risk may turn into an issue and be managed as an issue.

J

Job description: An outline of the skills, knowledge, responsibilities, and relationships involved in the work position. Also called the position description.

Job order: A written instruction by an authority for work to be done.

Joint venture: A contractual arrangement between two or more organizations to collaborate on a project to achieve common goals and mutual profit.

Just in time: A coordinated approach of resources, requirements, and production where the right material is available at the right place and time for use.

K

Kanban technique: A scheduling system to help decide what to produce, when to produce it, and how much to produce, a visual system for managing work as it moves through a process, involves lean and just-in-time production.

Key component of risk: A description of the negative or positive incident associated with risk.

Key event schedule: See Master schedule.

Key event schedule: See Milestone schedule.

Key event: See Milestone.

Key performance indicators (KPIs): Measurable indicators used to report progress selected to reflect the critical success factors of the project. KPIs evaluate the success of an organization or a project.

Kick-off meeting: A meeting at the start of the project or at the start of a significant phase of a project to officially start the project and help the team understand objectives, procedures, and plans.

Kill point: See Control gate.

Knowledge management: It refers to a multidisciplinary approach to achieving organizational objectives by making the best use of knowledge. The process of creating, sharing, utilizing, and managing the knowledge and information of an organization.

Knowledge transfer: Flow of knowledge and competencies from one person to another (For example, the project team members spending time with the end-user following a project implementation to show the users how to operate the new solutions).

L

Labor efficiency: The ratio of earned hours to actual hours spent on a project activity or task.

Labor hour contract: A fixed price type contract where a fixed amount is paid for each hour of work performed by a resource.

Labor rate variances: The difference between planned labor rates and actual labor rates.

Labor: The effort consumed by resources for pay or salary.

Ladder: A representation of overlapping activities in a network diagram (For example, the start and finish of each succeeding activity are linked only to the start and finish of the preceding activity by lead and lag activities).

Lag relationship: A reference to one of four possible network diagram activities relationships (1) start-to-start. (2) finish-to-start, (3) start-to-finish, and (4) finish-to- finish.

Lag: The time delay between the start or finish of activity and the start or finish of its successor.

Laissez-faire management style: The direct opposite of autocratic management style. Instead of the project manager making all decisions, laissez-faire leaders managers make few decisions and allow their team to choosing appropriate solutions. (For example, a highly skilled and mature team).

Late dates: A calculated backward pass time analysis, late dates are the latest dates by which an activity can be allowed to start or finish.

Late event date: The latest date an event can occur, calculated from the backward pass,

Late finish date: The latest possible date that activity may finish without delaying a specified milestone.

Late start date (LS): The latest possible date that activity may begin without delaying a specified milestone.

Lateral communication: The exchange or sharing of information between people within a community of peer groups (For example, communication with the same hierarchical level within an organization).

Latest event time: The latest time by which an event has to occur within the logical and imposed constraints of the network.

Latest finish time: The latest possible time by which activity has to finish within the logical activity and imposed constraints of the network.

Latest revised estimate: See Estimate at completion.

Latest start time: The latest possible time by which activity has to start within the logical and imposed constraints of the network.

Law of diminishing returns: The decrease in the marginal output of a production process as the amount of a single factor of production is incrementally increased, while the amounts of all other factors of production stay constant (For example, beyond a certain level, productivity increases at a decreasing rate, such that for every unit in efforts to increase productivity, expect less than a unit of productivity in return).

Lead contractor: The prime contractor assuming responsibility for overall project management and quality assurance. See also, Prime contractor.

Lead time: The time required to wait for the availability of product, service, or resources following its confirmed order.

Lead: The minimum time between the start of one activity and the start of an overlapping activity in a network diagram (For example, in a finish-to-start dependency with a 3-day lead, the successor activity can start 3 days before the predecessor has finished.

Leader: An individual with the capacity and skills to inspire and influence people (For example, possess the skills such as Honesty, integrity, Confidence,

Commitment, Passion, communication, and Decision-Making Capabilities).

Leadership: The skill and ability to establish a vision, direction and influence others towards a common purpose, and to empower and inspire people to achieve project success.

Learning and development: Involve the continual improvement of competencies in the organization. The identification and application of learning within projects develop the organization's capability to undertake current and future projects.

Learning curve theory: A parametric model stating that effort to complete a task should take less time and effort the more the same task is done over time (For example, the more times you do a task, the less time it takes to complete the same task).

Legal awareness: The level of understanding project management professionals of the relevant legal duties, rights, and processes that should be applied to projects.

Legitimate authority: See Formal authority.

Lessons learned: A vital process improvement documentation of the captured lessons following the completion of a project (For example, statements describing what did or did not work well).

Letter of intent: A formal letter indicating intent to sign a contract so that work can commence prior to signing that contract (For example, an intent to sign a contract following internal process but authorizing the project manager to start mobilizing the team).

Level finish schedule: The date in which an activity is scheduled to complete using the resource allocation process.

Level of approval: The seniority level assigned by management for approval (For example, the CEO is authorized to approve on behalf of the organization).

Level of effort (LOE): A formal document describing what is lawfully acceptable in accordance with common practices, activities that are not measured by accomplishment but rather a uniform rate of activity over a specific time period.

Level of effort contract: A type of contract stating the amount of effort required (For example, man-hour, man-week, man-month).

Level start schedule: The date in which an activity is scheduled to begin that is equal to or later in time than its start.

Leveling: See Resource leveling.

LF: Late Finish date

Liability period: The time during which a prime contractor or subcontractor is liable for the failure of a delivered product.

License: An official permission needed to do a particular exercise or a specific privilege (For example, licenses may be granted by governments, businesses, or individuals).

Licensing agreement: A contractual arrangement in which an organization that owns a particular technology authorizes its use by another organization under certain conditions.

Life cycle: A sequence of defined stages over the duration of a project.

Likelihood: Assessment of the probability that a risk will occur.

Limitation of cost clause: An essential clause in a cost-reimbursement contract to mandate that the contractor use best efforts to perform the agreed project tasks within the estimated cost.

Limitation of funds: The total funds available to the project manager within a period of time or for the entire project. Once the funds are consumed, no work will be authorized till additional funds are available (For example, in a cost- reimbursement contracts where the contractor must inform the buyer of the need for additional funds to continue or complete project activities).

Limiting quality: The maximum number of acceptable product quality defects.

Line function: Part of the organization that is responsible for or performing specified project activities (For example, producing certain products or performing certain services).

Line item: An item appearing on a single line whose status is tracked in a status system and which is usually a deliverable (For example, An item of revenue or expenditure in a budget or other financial statement or report).

Line manager: A manager responsible for the activities of a particular line of business or functional department. (For example, engineering, manufacturing, marketing). See also, Line manager.

Line of balance: A management control process for collecting, measuring, and presenting facts relating to time, cost, and accomplishment measured against a specific plan (For example, a graphical representation of a production system of material from raw to assembly to shipment).

Link A relationship between two or more activities or tasks. See Logical relationship.

Linked activity: An activity that is dependent on the performance of another activity in precedence diagramming.

Linked bar chart: A bar chart that shows the dependency links between activities

Liquidated damages: Damages amount that the parties assign during the development of a contract for the injured party to be paid as compensation upon a specific breach.

Loading factor: A ratio of scheduling allowance for administrative work and project rework.

LOE: See Level of effort.

Log Frame: A tool used for improving the planning, implementation, monitoring, and assessment of projects, a way of structuring the main elements in a project and highlighting the logical linkage between them.

Logic diagram: a diagram that displays the logical relationships between project activities. See Network diagram.

Logic: See Network logic.

Logical relationship: A dependency relationship between two or more project tasks or between tasks and milestones, such that one cannot start or finish before another has started or finished. See also precedence relationship.

Loop: A network path that passes the same node twice.

Lose-lose result: Outcome of a conflict that results in both parties being worse off than when the conflict started.

Loss: Failure to make the project profitable (For example, a situation where the costs of performing a project have exceeded the amount paid to the contractor under the contract).

Lot formation: The process of segregating production units by identifiable groups (For example, by type, grade, or size).

Lot size: The number of product units that create a lot.

Lot: A group or collection of product units identified and treated as a single entity. Also called a batch.

Lowest overall cost: The least expenditure of funds over the lifecycle of a project counting all cost items into consideration.

LS: See Late start date.

M

Maintainability: The probability of performing a successful repair action within a given time (For example, the ease and speed with which a system can be restored to operational status after a failure occurs).

Maintenance guarantee: The assurance given by the product provider that it will be maintained during a specified period of time.

Maintenance quality assurance: A determination that product maintained or modified conforms to the conforms with technical requirements.

Major defect: A product defect that will probably result in failure or dramatically decrease its usability.

Make or buy analysis: A management practice of analyzing the possibilities, advantages, and profit vs. disadvantages and cost to determine if needed product or service can be produced or performed within the organization should be purchased or subcontracted to another organization.

Management by exception: The practice of examining the financial and operational results of a project, and only bringing issues to the attention of management if results represent substantial differences from the budgeted.

Management by objectives (MBO): A strategic management model that aims to improve the performance of an organization by clearly defining objectives that are agreed to by both management and employees.

Management by projects: A management approach used to treat operation aspects in an organization as projects (For example, applying project management principles and practices to ongoing operations.

Management development: Organizational aspects of staff planning, recruitment, development, training, and performance assessment.

Management reserve: A separately planned amount of budget to account for parts of the project that cannot be predicted. (For example, unknown unknowns, these are risks that are unknown to the project and maybe triggered without advance notice).

Mandate: A document that defines the goals and constraints of a project and functions as a working agreement or contract between the client and the project manager.

Mandatory dependency: Is one that "must be" carried out at a particular time. ... When the sequence of events is developed for various aspects of the process, mandatory dependencies are placed where they must happen. Also called hard logic.

Margin: Profit gained above the cost of the product or service delivered.

Marginal cost: The cost added by producing one additional unit of a product or service the increase or reduction in cost based on the result on a unit of output.

Market requirements: The set of requirements for a business environment in which a product or service is targeted to be sold.

Market survey: An organized effort to gather information about target markets or customers. It is an essential component of business strategy.

Master network: The network displaying the complete project, from which more detailed networks are derived

Master Schedule: A level schedule that identifies the significant activities and critical milestones. See also milestone schedule.

Material breach: A contractual term which refers to a failure of performance under the contract, which is significant enough to give the aggrieved party the right to sue for breach of contract.

Material requirements planning (MRP): Analyzing project material needs and ordering techniques based on project requirements to account for lead time for purchase or fabrication.

Material: Goods that may be incorporated into or attached to an end item to be delivered under a contract (For example, raw and processed material, parts, components, assemblies).

Materials management: An organizational process and regulations for the ordering, storage, and movement of materials.

Mathematical Analysis: See Network analysis.

Mathematical programming: A mathematical technique for solving constrained optimization problems.

Matrix organization: Any organization where the project manager shares responsibility with the functional managers for assigning priorities and for directing the work of resources assigned to the project.

Maturity level: A defined position of an organization on an achievement measurement scale that describes the organization's capabilities.

Maximum fee: The maximum amount in a cost-plus-incentive-fee contract that the contractor can receive.

MBO: See Management by objectives.

Mean downtime: The average time that a system is non-operational (For example, all downtime associated with the repair, corrective and preventive maintenance, self-imposed downtime, and any logistics or administrative delays).

Meantime between failure (MTBF): The predicted elapsed time between inherent failures of a mechanical or electronic system during regular system operation (For example, MTBF can be calculated as the arithmetic mean time between failures of a system).

Meantime to repair (MTTR): A basic measure of the maintainability of repairable items (For example, the average time required to repair a failed component or device).

Measurement of cost: The process of planning and controlling the budget of a project.

Measuring and test equipment: Devices used to measure and test equipment to determine whether their compliance with technical requirements.

Mediation: The process of bringing disputed parties together to settle their differences through a meeting with a neutral party, the mediator.

Megaproject: A vast project, usually valued at US$100 million or more.

Methodology: See project management methodology.

Methods and procedures: The details of the standard practices to be used for managing projects throughout a lifecycle. (For example, methods provide a consistent framework within which project management is performed, and procedures cover individual aspects of project management practice and form an integral part of a method.

Metrics: Units of measurement used to assess or determine progress performance in terms of monetary units, schedule, or quality results.

Mid-stage assessment: An assessment carried out in the middle of a project (For example, a request of the project board, to authorize work on the next stage before the current one is completed, or to allow for a formal review in the middle of a long project).

Midpoint pricing: The price between the best price of the sellers and the best price of the buyers' bid price (For example, the average of the current bid and ask prices being quoted).

Migration plan: A document to plan a move from one business environment to another (For example, migration from one software to another).

Milestone chart: A scheduling technique to show the start and completion of milestones on a time-scale chart. Also, see Event chart.

Milestone method: An approach of calculating an earned value based on a weighted value assigned to each milestone.

Milestone plan: A plan containing only milestones of the project.

Milestone schedule: A schedule showing key events or milestones (For example, key accomplishments planned at time intervals throughout the project). See also Master schedule or summary schedule).

Milestone schedule: a schedule that identifies the significant milestones. See also Master schedule.

Milestone: A milestone is an event; it has no duration or effort. It must be preceded by one or more tasks. A point in time when a deliverable or set of deliverables is complete.

Minimum fee: The minimum amount in a cost-plus-incentive-fee contract that the contractor can receive.

Minor defect: A product defect that may not result in failure or dramatically decrease its usability but still needs to be addressed.

Minor risk: Risk event that does not cause significant problems, regardless of its probability.

Mission statement: A brief summary, approximately one or two sentences, that sums up the background, purposes, and benefits of the project.

Mitigation Strategies: Identification of the steps that can be taken to lessen the risk by lowering the probability of a risk event's occurrence, or to reduce its effect should the risk event occur.

Mixed organization: An organizational structure that includes both functions and projects in its hierarchy.

Mobilization: The exercise of bringing together project resources to start project activities (For example, personnel, equipment, and facilities).

Mockup: A built to scale demonstration model of the actual product used early in the development of a project to verify a proposed design, fit for purpose and other physical characteristics of the end product).

Modern project management (MPM): A term used to distinguish the current broad range of project management (scope, cost, time, quality, risk, and all other aspects of project management.) from narrower, traditional use that focused on cost and time.

Modification: Any changes to project scope, schedule, or cost (For example, change orders or requests).

Moments of truth (MOT): Point in time when the project owner or user starts operating the project end product or service (For example, a judgment of the project's product or service).

Monte carlo analysis: A schedule risk assessment method that relies on repeated random sampling to obtain numerical results by performing a project simulation many times to calculate a likely distribution of results. See also Schedule simulations.

Most likely time: In PERT estimating, the most realistic number of work periods the activity will consume.

Motivating: Encouraging project team members to perform better.

Motivation hygiene theory: Theory of motivation proclaims that certain factors in the workplace cause job satisfaction while a separate set of factors cause dissatisfaction, all of which act independently of each other (For example, project managers focus on team devotion to the project performance and use recognition as motivators).

Motivators: Factors related to job satisfaction that must be addressed to motivate the project team (For example, recognition).

MPM: Modern Project Management

MRP: See Material requirements planning.

Multidisciplinary: Combining or involving several project disciplines or professional specializations in an approach to a or problem.

Multi-project analysis: Analysis of the impact and interaction of activities and resources whose progress affects the progress of projects with shared resources or both (For example, collective reporting on projects having dependencies or resources in common).

Multi-project management: Managing multiple projects that are interconnected logically or by shared resources.

Multi-project scheduling: The use of resource allocation to schedule more than one project concurrently.

Multi-project: A project consisting of multiple sub-projects.

Murphy's law: An adage that is typically stated as: "Anything that can go wrong will go wrong."

N

Natural variation: The statistical description of natural fluctuations in process outputs, A quality concept stating that making any product with absolute consistency is very difficult.

Near-critical activity: An activity that has low total float.

Negative cash flow: A situation in which a project is spending more than it is earning.

Negative float: An indication that activities must start before their predecessors finish in order to meet a target finish date (For example, when the difference between the late dates and the early dates (start or finish) of activity are negative).

Negative total float: The time by which the duration of an activity or path has to be reduced in order to accommodate a limiting imposed date to be achieved.

Negotiated bidding rates: Agreed rates with the customer based on a reasonable estimate of the cost

Negotiated contract cost: The estimated cost agreed in a cost-plus-fixed-fee contract or the agreed contract target cost. See also Contract target cost.

Negotiation: A search for agreement, seeking acceptance, consensus, and alignment of views. Negotiation in a project can take place on an informal basis throughout the project lifecycle or on a formal basis such as during procurement, and between signatories to a contract.

Netbook value: The amount shown for assets, liabilities, or equity.

Net loss: A condition in which the expenses for an accounting period exceed income.

Net present value (NPV): The value in the present of a sum of money, compared to a future value it will have when invested at compound interest.

Net profit: An earned amount of money after all direct and indirect expenses have been deducted from total revenue.

Network analysis: The process of identifying early, late start and early, late finish dates for project activities.

Network-based scheduling: A determination of logical relationships between project WBS work packages, activities, and tasks to establish the shortest possible project duration (For example, PERT, CPM, and PDM).

Network diagram: A graphic tool for displaying the sequence and relationships between tasks in a project (For example, PERT Diagram, Critical Path Diagram, Arrow Diagram, Precedence Diagram, are all forms of network diagrams).

Network interface: Activity or event common to two or more network diagrams.

Network Logic: The collection of activity dependencies making up a project network diagram.

Network Path: A continuous series of connected activities in a project network diagram.

Network: A graphical presentation of project activities and tasks in which the project logic is the primary determinant of the placements of the activities in the drawing. See Project network diagram.

NGO: A non-governmental organization is a non-profit organization (For example, NGOs often conduct humanitarian and development work around the world.

No earlier than: A restriction on an activity to suggest that it may not start or end earlier than a specific date.

No later than: A restriction on an activity to suggest that it may not start or end later than a specific date.

Node: One of the defining points of a network; a junction point joined to some or all of the other dependency lines. See also arrow diagramming method and precedence diagramming method.

Nodes: points in a network where arrows start and finish.

Nominal group technique: A technique for small group discussion in which ideas and requirements are ranked and prioritized by all the members of the group after generation of all the ideas and requirements.

Nominal rate: An unadjusted for an inflation rate of return on investment.

Non-fragmented activity: An activity that, once started, has to be completed to plan without interruption.

Non-recurring costs: Expenditures against specific occur only once on a given project.

Nonbinding arbitration: Arbitration where the parties are not legally bound to the decision of the arbitrator. See also Arbitration.

Nonconformance: Failure to conform to accepted standards, A deficiency that makes the quality of goods or service unacceptable.

Nondisclosure agreement (NDA): A contract by which one or more parties agree not to disclose confidential information that they have shared with each other as a necessary part of doing business.

None Tangible Asst: Assets that may not have a physical presence like goodwill and brand recognition.

Nonpersonal services contract: A contract where the personnel rendering the services are not subject either by the contract's terms or by the manner of its administration, to the supervision and control usually prevailing in relationships between the Government and its employees

Nonwork calendar unit: A project calendar unit in which project work will not be performed (For example, national holiday).

Norming: This occurs as individuals become less defensive and more interested in finding solutions. A willingness to listen and accept other views. A sense of group identity begins to form as the team realizes that it can find ways to work together to get the job done. An appropriate leadership style at this stage involves providing support, coaching, and recognizing people's efforts.

NPV: See Net present value.

O

Objective quality evidence: A statement of fact, quantitative or qualitative, relating to the quality of a service or a product based on observations, measurements, or tests that can be verified.

Objective: The set of desired outcomes of the project or any part of the project, both in terms of concrete deliverables and behavioral outcomes (For example, improved service, more income).

OBS: See Organizational breakdown structure.

Off the shelf item: An item manufactured, produced, or assembled and placed in stock before orders or contracts are received for its sale.

Offer: Response to a request that would obligate the contractor to supply the products or perform the services if selected.

On hold: Status describing a change from Active to being held (For example, due to other work demands or priorities).

Open delivery contract: A contractual arrangement in which the period of project delivery is not specified but is established during a performance.

Open door policy: A management approach that encourages employees to approach, speak freely, and regularly to management regarding any aspect of the business or project.

Open-ended problem: A problem with no identified single correct answer or boundaries.

Open quantity contract A contractual arrangement with an unspecified quantity of goods where deliveries are scheduled by placing orders with the contractor.

Operating characteristic curve: A chart that displays the probability of acceptance versus a percentage of defective items (or lots).

Operation phase: The phase when the completed deliverable is used and maintained in service for its intended purpose.

Operations and maintenance (O&M): The phase that usually follows project acceptance (For example, O&M is not part of the project lifecycle but is part of the product lifecycle).

Opportunity assessment: Investigation of the possible occurrence of an event that is expected to have a positive impact on a project.

Opportunity cost: The loss of further alternatives when one alternative is chosen (For example, the rate of return that would have been earned by selecting an alternative project rather than the one selected.

Opportunity: A future event to enhance the project benefits (For example, a risk could be positive and provides for an opportunity).

Optimistic time: The minimum expected number of work units the activity will consume.

Order of magnitude estimate: A none detailed estimate carried out to give an approximate indication of costs. (For example, the order of magnitude estimate is within -25% to +75%). See Feasibility estimate.

Organization chart: A graphic display of the organization's hierarchy and reporting relationships.

Organization cost: Cost related to organization or reorganization expenses (For example, business license, legal fees, and consulting fees).

Organization Structure: The organizational environment within which the project takes place, it defines the reporting and decision-making order of an organization and how project management operates within it.

Organization: A public or private enterprise, firm, or company, either public or private.

Organizational breakdown structure (OBS): A hierarchical approach in which the organization may be divided into management levels and groups for planning and control purposes and to relate work packages to organizational units.

Organizational Culture: The behavior and unspoken rules shared by people in an organization (For example, the underlying beliefs, assumptions, values, and ways of interacting that contribute to the unique social and environment of an organization that is developed over time).

Organizational interfaces: The reporting and working relationships among different organizational units and divisions (For example, formal and informal relationships).

Organizational Planning: Identifying, documenting, and assigning project roles, responsibilities, and reporting relationships.

Organizational resources: Human and other resources available to the organization to achieve its objectives.

Organizational roles: The roles performed by individuals or organizational units in a project (For example, roles and responsibilities must be defined in to ensure clear accountabilities.

Organizational strategy: The sum of the actions a company intends to take to achieve its goals, these actions make up a strategic plan and Strategic requiring involvement at all company levels

Original budget: The initial budget established at the time a project was authorized, based on the negotiated contract cost or management's authorization.

Original duration: The duration of activities or groups of activities, as documented in the baseline schedule.

Original inspection: First inspection of a product compared to the inspection of a product that has been resubmitted following a rejection.

Out of sequence progress: Progress achieved although activities that have been deemed predecessors in the project have not been completed.

Outcome: The intended or achieved consequence of an output requiring the collective effort of relevant stakeholders.

Output format: The presentation of information that governs the final appearance of a report or drawing.

Output: The deliverable items that are a result of a process.

Outsourcing: The process of awarding part of or a complete project contract to a third party to perform services that may be performed by an organization's resources.

Overall change control: The coordinating and controlling of changes across an entire project (For example, changes related to scope, schedule, or budget, usually a change in either aspect of a project has multiple effects on other aspects as well).

Overhead: Costs incurred as a result of business operation that cannot be directly related to individual products or projects. See also Indirect cost.

Overlap: See Lead.

Overrun: The amount of the cost incurred is higher than the amount estimated.

Overtime: Excess time spent by project resources beyond the agreed regular working hours.

Ownership of quality responsibility: The responsibility of the project group or individual performing a task to ensure quality requirements are met.

P

Pairwise comparison: Any process of comparing entities in pairs to judge which of each entity is preferred or has a more considerable amount of some quantitative property, or whether or not the two entities are identical (for example, requirements or specifications).

Parallel activities: Two or more activities than can be done at the same time (For example, parallel activities could allow a project to be completed faster than if the activities were performed serially).

Parallel tasks: Two or more tasks than can be done at the same time

Parametric estimating: An estimating technique where parameters that represent different attributes of the project are used to calculate project effort, cost, and/or duration. (For example, square footage in construction, lines of code in software development) to calculate an estimate.

Parametric modeling: Project parameters expressed in a mathematical model to predict project costs.

Parent activity: A task within the work breakdown structure that includes subordinate tasks called child tasks.

Parent organization: An organization with holding power or authority of other organizations.

Pareto Diagram: A histogram displayed by frequency of occurrence, that shows how many results were generated by each identified cause.

Pareto optimal solution: A state in which making one party better requires making another party worse (For example, improving one of the functions while degrading some of the other functions).

Pareto's law: Also known as the 80/20 rule, is a theory asserting that 80 percent of the output from a given situation or system is determined by 20 percent of the input. The principle doesn't stipulate that all situations will demonstrate that precise ratio – it refers to a typical distribution.

Partial audit: A limited scope audit of the project or part of it.

Partial payment: A part payment that is made upon completion of a project milestone or delivery and acceptance of one or more completed deliverables.

Participative estimating: An estimating approach where other internal or external stakeholders participate in the estimate along with the primary estimator.

Participative management style: A management approach where the project manager asks input from team members and shares decision-making authority with them.

Parties to a contract: The persons or organizations who sign a contract with one another.

Partnership: A structured working relationship of two or more parties to achieve common business goals.

Party of interest: See Project stakeholder.

Patch deliverables: Project deliverables for project requirements that are developed as they occur (For example, a project that lacks comprehensive capture of all its requirements, may also be associated with time and material arrangement). See also Time and material arrangement.

Path convergence: The point at which parallel series activities paths meet (For example, the tendency to delay the completion of a milestone where the paths meet.

Path Float: See Float.

Path: A set of successively connected activities in a project network diagram.

Pay for performance: A compensation structure in which employees of an organization are compensated based on how their performance is assessed (For example, organizations adopt this approach to improve and enhance performance).

Pay period: The period of time used to determine the pay earned by an employee (For example, two weeks, or monthly).

Payback period: A project financial evaluation criterion to determine the length of time before the net cash flow from a project becomes positive.

Payment authorization: The process of allocating the required funds to pay a contractor for delivered products or services in accordance with the contract's terms and conditions.

Payment bond: A pledge to secure payment by the prime contractor.

Payment: Commitment to reimburse a contractor or an organization for work performed in accordance with the contract's terms and conditions.

PC: Percent complete

PDM: Precedence diagramming method

PDR: See Preliminary design review.

Peer audit: Audit conducted by a group of peers rather than full-time or external auditors (For example, multiple teams working on different parts of the project may audit their peers).

Peer review: A review of a project or phase of a project by individuals who are not part of the project team but with equivalent knowledge and background.

Pending: Status describing project work submitted for review but not yet reviewed for approval.

Percent Complete (PC): A percentage estimate of the work that has been completed on an activity or a project.

Percent complete: The percentage of the project completion that has been achieved.

Percent defective: The rate of defective units of a given quantity expressed in percentage.

Performance bond: See Bid bond.

Performance improvement: Performance improvement resulting from project team skills and behavior development.

Performance Indicators: See Key performance indicators.

Performance measurement baseline: a time-phased resourced plan against which the accomplishment of authorized work can be measured. It includes the budgets assigned to scheduled control accounts and the applicable indirect budgets.

Performance measurement techniques: The methods used to estimate earned value (For example, different methods are appropriate for different work packages due to the nature of the work or the planned duration of the work).

Performance Outcomes: Results of the efforts and actions during the project's planning and execution.

Performance Reporting: The exercise of collecting and distributing information about project performance to ensure project progress.

Performance review: Periodic project team meetings to assess project status or progress.

Performance specification: A description of the operational and technical characteristics of a product, together with its expected quality and performance.

Performance: A measure of project quality achievement.

Performing Organization: The organization responsible for performing the work in a project.

Performing: The stage that everyone waits for. The team finally reaches its optimum level of performance. Team members become focused on the common goal and work with a more collaborative style. Conflicts are less destructive than those of the storming stage, and cooperation is preferred. By this stage, the group is ready to get on with the job at hand.

Personal services contract: A contract under which the personnel rendering the services are referenced, either by the contract's terms or by the manner of its administration (For example, employees of the buyer's organization). See Nonpersonal services contract.

PERT: See Program evaluation and review technique.

Pessimistic duration: The maximum number of work periods the activity will consume. Also called pessimistic time.

Pessimistic time: See Pessimistic duration.

PF: Planned finish date

Phase (Stage) gate: See Control gate.

Phase end review: See control gate.

Phase exit: See Control gate.

Phase: See the Project phase.

Phased delivery approach: A delivery of a project in phases to allow end-users to utilize certain functionalities rather than wait until the entire project is delivered or the total project to be completed.

Phased project planning: A project planning method that involves project planning based on the available information.

Pilot: A test to establishes the implementation of a new product or solution prior to committing to its full release.

Plan: An intended future course of action.

Planned activity: Activity or task not started as of the current date.

Planned cost: A cost estimate of achieving an objective.

Planned finish date (PF): See Scheduled finish date.

Planned start date (PS): See Scheduled start date.

Planned value (PV): The total budgets for all planned project work scheduled to be accomplished within a given time period. See also Budgeted costs of work scheduled (BCWS).

Planner: A project team member project management office assigned the responsibility for planning, scheduling, and tracking of projects.

Planning phase: See Development phase.

Planning processes: All activities associated with developing and maintaining a workable project plan to accomplish the project objectives.

Planning: The process of identifying the method, resources, and actions necessary to accomplish a project.

Platform: The infrastructure setup of the software, hardware on which applications are developed, installed, and operated.

Plug date: An externally assigned date to an activity which establishes the earliest or latest date the activity is allowed to start or finish.

PM: Project Management or Project Manager

PMBOK®: Project management body of knowledge

PMO: See Program management office.

PMP®: Project management professional

Point of contact (POC): A person or an organizational entity serving as the coordinator or focal point of information concerning activity or program (For example, they are used in WHOIS contacts).

Policy deployment: The process of developing and disseminating policies in the organization (For example, interpreting senior management's objectives and directions into more precise and quantifiable objectives for each unit in the organization.

Portal: A web site developed to provides services, information, and references on project management. See also Project management portal

Portfolio management: The management of a collection of projects and programs as a single unit due to their relationship or common strategic objectives.

Portfolio: A collection of projects, programs, and operations managed as a group to achieve strategic objectives and to allow prioritization based on factors such as alignment with corporate strategy, risk, and technology focus.

Position description: See Job description.

Positional negotiation: A negotiation strategy that involves holding on to a fixed position and arguing for it and it alone, regardless of any underlying interests.

Positive float: The amount of time that an activity's start can be delayed without affecting the project completion date (For example, when the difference between early and late dates (start or finish) determines the amount of float.

Post contract evaluation: A performance review of all contractual parties following the conclusion of the contract to

Post-implementation review (PIR): An evaluation following project completion by the project team to learn from the successes and failures experienced during the project. It determines if the expectations established for the project outcome were met and makes a comparison between the actual results of a project's objectives and deliverables specified in the project's Charter.

Post project appraisal: A project team performance evaluation intended to benefit team members for future projects.

Post project evaluation and review: See Lessons learned review.

Power: The ability to influence the actions of others (For example, formal delegation of authority, reference power, or subject matter expertise).

Pre-award meetings: Meetings with prospective contractors before the final contract award.

Pre-award survey: An evaluation of a potential contractor's ability to perform under the terms and conditions of a contract.

Pre-commissioning: The work that is carried out prior to project commissioning to demonstrate the safety of the undertaken.

Precedence diagramming method (PDM): Also called activity on node (AON) or activity on arc. A network diagramming technique where activities are represented by nodes, nodes connected with arrows to show dependencies, activities are linked by precedence relationships to show the sequence in which the activities are to be performed, and the relationship between them is linked by arrows. (For example, only types of relationships possible are: start to start, start to finish, finish to start and finish to finish).

Precedence Relationship: The term used in the precedence diagramming method for a logical relationship.

Preceding event: In an activity on arrow network, an event at the beginning of an activity

Precontract cost: Cost sustained by a contractor before the contract's effective date.

Predecessor Task: A task (or activity) that must be started or finished before another task or milestone can be performed.

Prefeasibility phase: See Concept phase.

Preferential logic: See Discretionary dependency.

Preferred logic: See Discretionary dependency.

Preliminary design review (PDR): A technical assessment of all configuration items to evaluate the selected design approach for adequacy, compatibility, and requirements to ensure compatibility and functional interfaces (For example, a technical assessment that establishes the Allocated Baseline of a system to ensure a system is operationally effective).

Preliminary estimate: See order of magnitude estimate.

Premature termination: A situation where a project is terminated prior to achieving its objectives (For example, a decision by management due to pressing circumstances or a decision by the project owner to end the project).

Prequalification: A process to determine the qualification and readiness of potential contractors prior to the request for a proposal.

Present value: The monetary value in current units for work to be performed.

Price adjustment: A change of the price in a contract.

Price analysis: Investigation and evaluation of the project related price with no cost performance analysis (for example, a determination of the reasonableness of the offered price in comparison to other vendors offering the same product or service).

Price competition: A pricing strategy in which the competitors' prices are taken into consideration when setting the price of the same or similar product (For example, the focus is on competition-driven prices rather than production costs and overheads

Price: The payment or compensation given by a party to another party in return for a unit of goods or services

Pricing arrangement: The basis on which a project owner agrees to pay a contractor (For example, cost reimbursement arrangement or a fixed price arrangement).

Primary failure: A significant failure of a system that causes the system or application to be rendered nonoperational.

Prime contractor: See Lead contractor.

PRINCE2: See Projects in controlled environments.

Principled negotiation: An approach to negotiation with an objective of the achievement of a win-win result.

Priorities: An exercise of scheduling activities with previously imposed constraints.

Priority rules: A method used to rank items to determine which one should be first.

Privatization: The transfer of ownership from the government to the private where the government ceases to be the owner of the entity or business.

Privity of contract: The relationship and responsibilities between parties to the same contract.

Probabilistic network: network containing alternative paths where probabilities are associated.

Probability analysis: A technique used in risk analysis for forecasting future events, it involves a review of historical project loss information to calculate a probability distribution that can be used to predict future losses (For example, accidental and business losses).

Probability of acceptance: Describes the chance of accepting a particular lot based on a specific sampling plan and incoming proportion defective.

Probability: The likelihood of occurrence. It is usually associated with the likelihood of risk occurrence.

Problem definition: A statement on an area of concern, a condition to be improved, a difficulty to be eliminated, or a troubling question that exists that points to the need for meaningful understanding and deliberate investigation to determine the scope of the project.

Problem resolution: A discussion or conference between project stakeholders for a reason for finding a mutually acceptable solution to a project problem.

Problem-solving: Also called confrontation. See conflict management.

Problem statement: A formal document that defines the problem to be resolved (For example, a condition to be enhanced, a difficulty to be eliminated, or a troubling question that exists).

Procedures: A series of step by step actions conducted in a particular order to be performed (For example, a prescribed method to perform project work).

Process adjustment: The set of corrective or preventive action as a result of quality control measurement. See Quality control measurement.

Process analysis: A structured method to identify and understand how an organization works (For example, defines business processes and data flow through specific diagramming technique).

Process definition: An exercise of breaking down a process into its smallest components to describe it in detail.

Process flowchart: A diagram displaying the relationship of all elements of a system. Also called a system flowchart.

Process: A series of steps or actions to accomplish something. A set of interrelated resources and activities which transform inputs into outputs

Procurement management plan: Part of the overall project management plan that describes the management of the procurement processes, from solicitation planning through contract close.

Procurement Planning: Determining what to procure and when.

Procurement ranking: A measure of the competency, capabilities, and qualifications of potential contractors (For example, qualitative or quantitative measure for the purpose of selecting appropriate contractors to provide the required products or services).

Procurement strategy: The plan to efficiently acquire the necessary supplies from a list of competent vendors that will deliver quality goods on time and in accordance with contractual terms and conditions.

Procurement: The process by which the resources, goods, and services required by a project are acquired.

Product analysis: A process of developing a better understanding of the product of the project. It involves examining product features, costs, availability, quality, and others.

Product breakdown structure (PBS): A hierarchy of deliverable products involves the decomposition of a product into essential parts to provide a guide for configuration control documentation (For example, a Bill of quantities or materials).

Product description: The description of the components of a product (For example, features, characteristics, functions, and use).

Product development process: The documented and structured approach an organization follows in designing, developing, and introducing new products to the marketplace (For example, product development stages include concept, selection, design, development, testing, availability, and maintenance).

Product flow diagram: A representation of the order by which a sequence of products is created according to planning principles, it is related to the product breakdown structure.

135

Product liability: A reference to a manufacturer or a vendor being held liable for placing a faulty product into the hands of a consumer. Responsibility for a product defect that causes injury lies with all sellers of the product who are in the distribution chain.

Product lifecycle: The total time that a product exists from concept to termination.

Product quality review: The measurements and actions to determine the quality of delivered products comply with specified requirements.

Product scope: The collective characteristics, features, and functions to be included in a product.

Product substitution: A replacement of a product delivery that may or may not meet contractual requirements.

Product: The agreed tangible outcome of a project or project activities that are made available for use.

Production permit: A formal authorization to depart from a requirement or specification for a specified item and time.

Production readiness: A measure of the system readiness to proceed to the production phase.

Productivity: Efficiency and effectiveness measurement criteria (For example, labor efficiency or equipment effectiveness).

Profit and loss statement: A formally documented summary of all revenues and costs for a specified period.

Profit: See Net profit

Program benefits review: A review to measure if the project stated targets have been reached as well as measuring the performance levels in the resulting business operations.

Program director: The senior manager with the responsibility for the overall success of the program.

Program directorate: A committee that directs the program when there is no individual to direct the program.

Program Drivers: An explanation of why the project is needed and why it is being recommended at this time (For example, a description of the problem or issue that will be resolved by the project and any background information necessary to understand the problem).

Program evaluation and review technique (PERT): An event-oriented network study method used to estimate project duration when there is a high degree of uncertainty with the individual activity duration estimates.

Program management office (PMO): The office responsible for the business and technical management of a project or program, an organizational function to provides support on governance, project management best practices, mentoring, tools, and standardized processes. EPMO is an enterprise-wide function, while a PMO could be a departmental or organizational function.

Program management: the effective management of several individuals but related projects or functional activities in order to produce an overall system that works effectively.

Program manager: The person who directs the planning and execution of a program and is held accountable for the success of the program.

Program office: See Program management office.

Program support office: A function responsible for providing administrative support to the program manager and the program team.

Program: A group of projects selected, planned, and managed in a coordinated way to achieve a set of defined objectives; a single, large or very complex project with phases managed as separate projects; or a set of unrelated projects.

Progress payment: A payments made to a contractor during the life of a fixed-price type contract, on the basis of an agreed formula (For example, budget cost of work performed, or costs incurred).

Progress report: A regular report to project team members, project sponsors, or stakeholders summarizing the progress of a project, including crucial events, milestones, costs, and other issues.

Progress trend: An indication of the project progression rate (For example, is the project progressing positively, negatively, or remaining stagnant).

Progress: The partial measured completion of a project.

Project accounting: The process of identifying, collecting, measuring, recording, and communicating actual project financials.

Project appraisal: The discipline of measuring the feasibility and value of a project.

Project approach statement: A statement of the method the project will accomplish its objectives (For example, project planning methodology, risk management approach, and quality management approach).

Project archives: The complete documentation of project records (For example, records could be organized in historical or by an agreed specific method).

Project Assumptions: See Assumptions

Project audit: See Audit, Internal, audit, Partial audit, Full audit, and External audit

Project base date: A reference date used as a basis for the start of a project calendar.

Project baseline: See Baseline

Project board: A committee of stakeholders to which the project manager is accountable for achieving the project objectives.

Project brief: A statement that briefly describes the objective, time, budget, and performance requirements for a project.

Project budget: The sum of monetary funds allocated to a project.

Project business case: Essential information and documented analysis needed to drive approval, authorization, or assess a project proposal and reach a reasoned decision. See also Project business case.

Project calendar: A project calendar lists time intervals in which activities or resources can or cannot be scheduled (For example, Monday to Friday or Sunday to Thursday).

Project champion: An individual within the organization who promotes and defends a project.

Project change control board (PCCB): See Change control board.

Project change control: See Change control.

Project Charter: A formal document established by senior management to sanction the project and authorize the project manager to carry it out within the scope, quality, time, cost, and resources.

Project closure: The final phase of a project lifecycle, a formal termination that includes administrative and financial closure and release of the project team.

Project communication management: See Communication management.

Project Context: A reference to the environment within which a project is undertaken, an appreciation of the context where the project is being performed will assist project stakeholders involved in delivering a project.

Project coordination: Synchronization of the various project areas to ensure the transfer of information at interface points and appropriate times.

Project coordinator: An individual usually reporting to the project manager in charge of project coordination. However, in some organizations, such individual reports to a higher-level manager in the organization.

Project cost estimate: The sum of all costs required to complete a project.

Project cost management: See Cost management.

Project cost system: The cost accounting system that includes all project cost items (For example, assets, liabilities, expense, raw materials, and salaries).

Project culture: The general attitude toward projects within an organization.

Project definition: A document defining what the project requirements are, what will be done, how and the resources required to achieve its objectives.

Project direct labor: Labor applied to meet project objectives (For example, labor that can be used as an element in costing, pricing, and profit calculations)

Project director: See Program director

Project Donor The funding entity or organization that provides funds to the project.

Project duration: Total duration from the project start date to the project finish date.

Project environment: The context within which the project is formulated that includes all internal and external factors that have an impact on the project.

Project evaluation: See evaluation.

Project execution: See Implementation phase.

Project file: A file containing the overall plans of a project and all other relevant project documents.

Project financing and funding: The means by which the budget to commence a project is initially available and then made accessible at the appropriate time. (For example, projects may be financed externally, funded internally, or a combination of both). See Also Financing.

Project finish date: See Project planned finish date.

Project human resource management: See Human resource management.

Project indirect labor: Labor applied to overhead or general and administrative costs.

Project initiation document: a document approved by the project board at project initiation that defines the terms of reference for the project. See also Initiation.

Project integration management: See Integration management.

Project issue report: A formal report that raises either technical or managerial issues in a project.

Project justification: See Business case.

Project leader: See Leader.

Project lifecycle cost: The concept of including acquisition, operating, and disposal costs when evaluating various alternatives.

Project lifecycle: The collective sequential phases of project phases from a project's conception and its termination.

Project log: A chronological project record of all occurrences throughout the project

Project logic drawing: A representation of the logical relationships of a project.

Project logic: The relationships between all activities in a project.

Project management: The application of knowledge, skills, tools, and techniques to project activities in order to meet or exceed stakeholder needs and expectations from a project.

Project management body of knowledge (PMBOK®): An inclusive term describing the sum of knowledge within the profession of project management. The PMBOK includes proven, traditional practices which are widely applied as well as innovative and advanced ones which have seen more limited use.

Project management controls: The collectives project processes and procedures designed to ensure that performance information is collected, analyzed, and reviewed and used to achieve the project's objectives (For example, time management, scope management, and schedule management).

Project management information system (PMIS): The collection of information required for an organization to execute projects successfully (For example, software applications and the methodical process for collecting and using project information.

Project management institute (PMI®): International, nonprofit professional association dedicated to advancing the discipline of project management and state-of- the-art project management practices. See also Project management professional.

Project management lifecycle: The major sequential phases that any project passes, phases are time periods, namely: Initiation, Planning, Implementation, Monitoring, Closure, and each phase may be further broken down into stages.

Project management methodology: A highly detailed description of the procedures to be followed in a project lifecycle to ensure consistency (For example, templates, forms, charts, and checklists).

Project management office: The office or department responsible for establishing, maintaining, and enforcing project delivery and management processes, procedures, and standards. It provides services, support, and training for project managers.

Project management plan: A plan for carrying out a project to meet specific objectives that are prepared by or for the project manager. The project management plan is retained by the project manager.

Project management portal: A web site developed to provides services, information, and references on project management. See also portal.

Project management processes: A series of actions that describe and organize the activities and tasks of the project.

Project management professional (PMP®): An individual certified by the Project Management Institute, a professional certification awarded by the Project Management Institute to individuals who meet the requirements in knowledge, education, experience.

Project management software: Computer applications specifically designed to aid with planning and controlling project costs and schedules.

Project management team: Members and resources of the project team who are directly involved in project activities.

Project Management: Project management is the process by which projects are defined, planned, monitored, controlled, and delivered to realize the project objectives and goals.

Project Manager: The person responsible for project planning, managing, executing, and controlling the project and bringing the project in on time, cost, scope, and agreed on quality.

Project matrix: An organization matrix that is project-based, in which the functional structures are duplicated in each project.

Project metrics: See Metrics.

Project monitoring: The capture, analysis, and reporting of current project performance as compared to plan.

Project network diagram: See Network diagram

Project objectives: A description of the project's intended goals.

Project office: See Project management office

Project phase payment: A payment made to an individual or subcontractors at some predetermined milestone.

Project phase: A collection of logically related project activities concluding in the completion of a project deliverable.

Project plan: A formal, approved document used to direct project execution and control. The plan documents planning assumptions and decisions; it includes all other project plans such as Integration scope, cost, schedule, communication, resources, risk, quality, procurement, and stakeholder plans.

Project Planning: The process of the development and maintenance of the project plan.

Project politics: Actions necessary by a project manager or other stakeholder to influence a group of people with different interests to work toward the project objectives.

Project procedures manual: The standard set of management and administrative procedures needed for the project.

Project procurement management: Part of the overall project management that involves procurement planning, source selection, management of purchase orders, and contracts for products and services.

Project procurement management: See Procurement management.

Project progress report: See Progress report.

Project quality management: See Quality management.

Project requirements: See Functional requirements and also Technical requirements.

Project risk management: See Risk management.

Project risk manager: In large and complex projects, A person on the project team is assigned to the responsibility of all risk management (For example, preparing a risk management plan, executing and monitoring the plan).

Project risk: The collective positive or negative effect of the probability of uncertain occurrences that would affect project objectives.

Project schedule management: See Schedule management.

Project schedule: Prearranged dates and times intended for the performance of project activities.

Project scope management: See Scope management.

Project scope statement: A concise and accurate description of the expected work, products, and deliverables.

Project scope: All project work required to deliver a project's product or service in line with the project requirements.

Project segments: A subdivision of the project with its own work packages expressed as a manageable component.

Project selection methods: The organizational specific methods and used to select projects that best support the organization's strategy and objectives (For example, what benefits the organization most).

Project sponsor: The individual or organization with approval authority for a project to start and continue.

Project staffing requirements: a document detailing the skills needed for the project, how many and when they are needed.

Project stakeholder management: See Stakeholder management.

Project stakeholders: Individuals and organizations involved in a project or affected by project activities (For example, project managers, sponsors, and customers).

Project start-up: The creation of the project team.

Project Status Indicators: A project panel or a dashboard indicating the status of the project (For example, Green = On Track. Yellow = Caution, at risk. Red = Alert, a project in serious trouble).

Project status report: A report on the status of completion and any variances of the project.

Project strategy: See Strategy.

Project success - failure criteria: The agreed stakeholders' criteria by which the success or failure of a project may be judged.

Project Success: The satisfaction of stakeholder requirements as measured by the identified and agreed success criteria at the start of the project.

Project summary WBS: A higher-level WBS used for presentation and reporting to senior management or the project steering committee.

Project support office: See Program support office

Project team roles and responsibilities: Identified and agreed roles, responsibilities, and accountability of project team members individually and as a whole.

Project team: A group of individuals, groups and/or organizations that are responsible to the project manager for undertaking project tasks (For example, resources assigned to the project, contractors, and consultants).

Project technical plan: A plan produced at the beginning of a project that addresses technical and practical issues related to quality control and configuration management.

Project termination: See Closeout phase

Project time management: See Time management

Project: A temporary undertaking to create a unique product or service with a defined start and endpoint and specific objectives.

Projection: Estimate of project performance based on historical information, present situation, and future outlook.

Projectized organization: An organizational structure in which the project manager has full authority over project resources, assigns priorities, and directs the work of individuals assigned to the project.

Projects in controlled environments2: Project management method involving the organization, management, and control of projects, focus on a generic, best practice approach to project management.

Promotional management style: A management style where the project manager motivates and encourages project team members to realize their full potential (For example, by promoting team spirit and teamwork).

Proof of concept: Evidence to support acceptance of a proposed solution.

Proposal project plan: An initial plan that may be presented with a contractor's proposal (For example, the initial plan may include key milestones, deliverables, some fundamental analysis).

Prototype: A functioning smaller scale of a newly proposed product used to evaluate the end product design.

PS: Planned Start date

Public relations: An activity designed to improve the project organization's environment for the purpose of improving project performance and reception.

Punch list: A list of all project work outstanding or almost finished work to show the items of work remaining to fulfill the project scope.

Punitive damages: Monetary compensation, over and above actual damages, sought for nonperformance or wrongful acts.

Purchase description: A description of the characteristics necessary to meet project requirements.

Purchase order: An offer to buy certain products or services based on specified terms and conditions.

Purchase: An acquisition of products or goods (For example, an organization attempting to acquire goods or services to accomplish its goals).

Pure risk: See Insurable risk.

Purse string authority: Authority based on the amount of control an individual has over the project budget.

Q

QA: Quality assurance

QAR: See Quality assurance representative.

QC: Quality control

Qualitative risk assessment: A value-based assessment of a risk that includes the likelihood a risk will occur, its impact, response strategy, and ownership.

Quality Assurance (QA): The process of evaluating total project performance and making sure quality standards and procedures are practically complied with.

Quality assurance plan: Part of the quality management plan to ensure a quality approach and conformance of all requirements and activities in a project.

Quality assurance representative (QAR): An individual or organization with the sole responsibility of ensuring that the quality standards are applied and complied with.

Quality audit: A formal examination to determine whether practices conform to specific standards or a critical analysis of whether a deliverable meets quality standards.

Quality circles: An identified group within the project tasked with the process of products or services without management interference.

Quality conformance inspection: Examinations and tests performed on products or services to determine conformance with project requirements.

Quality control (QC): Making sure that project deliverables comply with agreed quality standards and acceptance criteria.

Quality control measurements: The results of quality control testing and measures for comparison and analysis.

Quality council: A committee of individuals within an organization responsible for coordinating the organization's quality program.

Quality criteria: The functionality and characteristics of a product that determines whether it meets specific standards.

Quality evaluation: A process of collecting project data for decision making in a quality process review.

Quality function deployment: A method developed to help transform the requirements of the customer into engineering characteristics for a product.

Quality gate: A predefined completion criteria for a project task (For example, project walk-throughs and inspections).

Quality guide: A document that describes quality and configuration management procedures aimed at people directly involved with quality reviews, configuration management, and technical exceptions.

Quality improvement: Action that the project manager and team take to increase the effectiveness and efficiency of the project.

Quality loop: A visual representation of the steps that interact in the process of creating a quality product (For example, the identification of a customer's needs and the eventual fulfillment of that need with the finished product. Also called quality spiral.

Quality management plan: A part of the overall project management plan, a document that describes how the quality standards, policies, and assurance will be implemented in a project.

Quality management: The process of planning, organizing, controlling, coordinating, and directing of project activities to achieve the stated project quality objectives.

Quality policy: The overall quality directions of the organization.

Quality risk: Failure to accomplish a project task in accordance with the required quality.

Quality spiral: See Quality loop.

Quality surveillance: An ongoing monitoring of quality system

Quality system review: A formal evaluation of the status and competence of the quality system.

Quality system: The organizational processes, procedures, tools, and resources utilized to implement quality management.

Quality: A project characteristic measure to assess the degree of excellence of a product or service.

Quantitative risk assessment: A numeric assessment of a risk that includes the likelihood a risk will occur, its impact, response strategy and ownership, and financial impact.

Quotation: See Bid.

R

RAM: Responsibility assignment matrix

Random cause: Unidentifiable reason that causes a special event that is outside the control limits.

Random sampling: A sampling method where each unit of the population has an equal chance of being selected.

Rayleigh curve: An equation specifying the relationship between applied effort and delivery time for a software project. In software project scheduling, the Putnam–Norden–Rayleigh curve known as the PNR curve.

RBS: See Resource breakdown structure.

RDU: Remaining Duration

Reasonable cost: The cost of a product or service in line with what would be incurred by others performing a competitive business.

Re-baselining: The exercise of establishing a new project baseline as a result of significant changes in the project.

Recession: Release of all obligations under a contract.

Records management: The organizational rules and procedures to identify, archive, restore, and distribute project documentation.

Recovery schedule: A particular type of project schedule used to show the effort required to recover time lost on a project compared to the master schedule.

Recruitment practices: The organizational policies, rules, and guidelines involving recruiting and assignment of staff.

Recurring costs: A repetitive cost that would occur against specific tasks (For example, hiring external resources, tools, and equipment).

Red team: In cybersecurity projects, a red team is offensive security professionals who are experts in attacking systems and breaking into defenses. See Blue Team

Referent power: a type of authority based on someone referring to a higher power to support their position (For example, using a reference like the CEO and I think this will work).

Regression analysis: A set of statistical processes for estimating the relationships between a dependent variable and one or more independent variables.

Regulation: A formal description of the characteristics and compliance requirements that a product or service need to adhere to.

Reimbursement: A payment to a party, a contractor, organization, or employee for incurred expenses in the project-related activity.

Rejection number: The minimum number of defective units in a sample that will cause the entire lot to be rejected.

Relationship: A logical connection between two activities.

Relative Date: A date expressed as a unit of time periods (For example, four days, two weeks, or a month).

Release: Agreement by a contracting party that the other party will not be liable for any future claims.

Reliability assurance: Agreed actions to satisfy conformance to approved reliability requirements.

Reliability: The quality of a product performing the agreed functions consistently.

Remaining available resources: The difference availability of resources and scheduled resources following the resource assignments to project activities and tasks.

Remaining duration (RDU): The time needed to complete an activity or a set of activities.

Remaining float (RF): The difference between the early finish and the late finish date of project activity.

Re-planning: Actions performed for any remaining effort within project scope (For example, cost, schedule variances are zeroed out at this time for history items).

Reporting: Communicating project information to stakeholders regarding project status and progress.

Representative sample: A subset of a population that seeks to accurately reflect the characteristics of the larger group (For example, a lot of 300 could generate a representative sample that might include 30 units).

Request for Applications (RFA): A formal request prepared for the development team to inform them of funds availability for the development work in a specific area.

Request for change: A proposal by the project manager or authorized stakeholder for a change to the project as a result of a project issue report.

Request for proposal (RFP): A formal request to solicit proposals from prospective sellers of products or services.

Request for quotation (RFQ): A type of request for a proposal but with more specific application areas.

Required approval: A request for a sign off to a higher authority.

Requirements analysis: A process of evaluating the stated needs and validating them against organizational requirements and plans.

Requirements definition: Statement of the needs that a project has to satisfy.

Requirements traceability: A requirements discipline that links requirements throughout the validation process. The purpose of the Requirements Traceability is to ensure that all requirements defined for a system are tested in an agreed test protocol.

Requirements: A negotiated set of measurable customer wants and needs.

Reschedule: A change in project activities start or finish dates or duration of an existing schedule (For example, as a result of corrective actions or imposed conditions).

Reserve: A financial provision in the project plan to mitigate the cost and/or schedule risk (For example, management reserve, contingency reserve set aside for unknown risks).

Residual risk: The amount of risk associated with an action or event remaining after an identified or risk has been reduced by risk.

Residual value: The remaining value of an asset following depreciation charges being subtracted from its original cost.

Resource accumulation: The process of collecting the requirements for each resource to provide the total required to date at all times throughout the project.

Resource aggregation: summation of the requirements for each resource, and for each time period. (For example, when the earliest start time of an activity is used alone, it is 'early start' aggregation, while a 'late start' aggregation uses the latest start times).

Resource allocation: The process of assigning resources to the activities in a network so that predetermined constraints of resource availability and/or project time are not exceeded.

Resource analysis: The process of analyzing and optimizing the use of resources on a project (For example, the use of resource leveling and resource smoothing techniques).

Resource assignment: The activities related to a specific resource.

Resource availability pool: Number and skills of project resources availability for a specified period.

Resource availability: The level of availability of a resource. See also Resource calendar

Resource breakdown structure (RBS): A hierarchical list of resources related by function and resource type that is used to facilitate planning and controlling of project work.

Resource breakdown structure: A hierarchical structure of resources to enable scheduling at a detailed level that rolls up to a higher level.

Resource calendar: A calendar that defines the working and non-working dates for specific resources.

Resource code: A distinctive resource identification method (For example, ID, skills, description, and any other related information).

Resource conflict: A situation where a resource is assigned to more than one project activity at the same time.

Resource-constrained scheduling: A resource-leveling situation where activities start and finish dates are calculated based on the availability of a fixed quantity of resources.

Resource constraint: The limitation of the availability of a resource.

Resource dependency: A dependency between tasks that share the same resources, therefore, cannot be performed simultaneously (For example, resource-dependent tasks can be scheduled at the same time but are limited by the availability of the shared resources).

Resource description: See Resource code

Resource driven task durations: A task duration that is driven by the need for a specific resource (For example, resource with a specific skill set and available for a specific time).

Resource histogram: A chart used to display resource utilization by time period. Also called resource loading chart.

Resource level: A specified level of resource units required by an activity per time unit.

Resource leveling: A form of network analysis where scheduling decisions are driven by resource management concerns (For example, limited resource availability or difficult to manage changes in resource levels).

Resource limited planning: Planning project activities taking into consideration that resource availability levels are not exceeded (For example, aligning resource availability and resource calendars with activity scheduling).

Resource limited schedule: A project schedule where the start and finish dates account for resource availability so predetermined resource levels are never exceeded (For example, activities start date is aligned to resource availability that may cause the specified project duration to be exceeded).

Resource loading chart: See Resource histogram.

Resource loading: The process of assigning resources to a project by activity (For example, people, facilities, and equipment).

Resource management plan: Part of the overall project management plan, a document that describes when and how human resources will perform activities in the project.

Resource management: The process of identifying, assigning resources to project activities to achieve project objectives with appropriate levels of resources and acceptable duration (For example, resource allocation, smoothing, leveling, and scheduling are techniques used to determine and manage appropriate levels of resources).

Resource matrix: A mechanism of assigning resources to tasks (For example, WBS tasks are listed on the vertical axis and the resources required on the horizontal).

Resource optimization: A reference to resource leveling and resource smoothing.

Resource planning: The process of determining the quantities of the resources needed to perform project activities.

Resource pool description: The aggregate information related to all project available resources (For example, people, equipment, material).

Resource rate: Unit rate for each resource required for the project (For example, (cost per hour or bulk cost).

Resource requirement: The requirement for resources with a specific skill set for a particular project activity.

Resource scheduling: The process of determining durations, dates, and times for resources to be available to perform project activities.

Resource selection: The process of selecting the type, number, skillset, and sources of resources necessary to accomplish project activities.

Resource smoothing: A technique that alters the activities of a schedule method so that requirements for the resources do not exceed the resource limits already pre-defined during the planning.

Resource spreadsheet: A tool used to show the number of resources, skill sets needed on the project, by type, duration, and dates.

Resource: Any personnel, material, or equipment required for the performance of project activity.

Resources: Human resources, material, and equipment required to accomplish project activities.

Response planning: The process of articulating response strategies to project risks (For example, avoidance of risk, acceptance of risk and the response strategy must include a risk owner to oversee the implementation of the response as well as monitoring and closing the risk once it is no longer posing a threat to the project).

Responsibility assignment matrix (RAM): A structure or a tool that involves relating each project activity in the WBS with a single resource or a functional organization responsible for completing the activity.

Responsibility chart: See Responsibility assignment matrix.

Responsibility: The obligation to perform a project task or activity with the liability to be accountable for loss or failure.

Responsible organization: a defined unit within the organization structure which is assigned responsibility for accomplishing specific tasks

Resubmitted lot: A resubmission of a lot that has been submitted for acceptance but rejected or removed.

Results based management (RBM): A management strategy by which an organization ensures that its processes, products, and services contribute to the achievement of desired results.

Retention of records: The period of time that records are kept for reference after a contract or project closeout.

Retention: A part of payment withheld until the project is completed in order to ensure satisfactory performance or completion of contract terms.

Return on equity (ROE): A percentage amount earned on investment for a given period of time.

Return on investment (ROI): A percentage rate earned on an organization's total capital.

Return on sales (ROS): A measure of operational efficiency that accounts for profits as a percentage of net sales.

Revenue cost: A total cost incurred to obtain a sale and the cost of the goods or services sold, it includes specific selling and marketing activities associated with a sale.

Revenue: The amount earned as a result of implementing a project (For example, the result of the project being a product or a service).

Reverse engineering: An inspection and analysis process by which a product is deconstructed to reveal its designs and architecture.

Reverse scheduling: A method where the project completion date is fixed; however, tasks and activities duration are used to determine the project start date.

Review: An examination to determine correctness or accuracy. See Phase end review

Revision: See Schedule revision

Reward and recognition system: A management mechanism to endorse or encourage desired positive behavior. (For example, rewards for completing tasks ahead of schedule, or reward for completing a project with less than cost than planned).

Reward authority: An authority bestowed to an individual to award project team members or other stakeholders to encourage positive behavior (For example, a project manager awards recognition or a bonus to top performers).

Rework: Steps and actions required to ensure that a defective or nonconforming item complies with requirements or specifications.

RF: See Remaining float.

RFP: Request for proposal

RFQ: Request for quotation

Risk acceptance: Part of risk response that involves preparation to deal with the outcome of risk. Developing a risk mitigation plan to execute if the risk event occurs is an active acceptance response.

Risk allowance: Budgeted money to cover project uncertainties due to estimates inaccuracies of risk events. See also Contingency reserve and Management reserve.

Risk Analysis: An examination of risk events to assess the probable consequences for each event, or combination of events and determining possible options for dealing with these events.

Risk appraisal: Effort involved in identifying and assessing risk.

Risk assessment: Part of the risk management plan that involves the process of identifying potential risks, quantifying their likelihood of occurrence, and assessing their likely impact on the project.

Risk avoidance: Part of risk response that involves planning activities to avoid risks that have been identified.

Risk budget: A combination of contingency and management reserves spent only if risks occur.

Risk consequence: See Impact.

Risk contingency: See Contingency.

Risk database: A database record of risks associated with a project.

Risk deflection: See Deflection.

Risk description: A description of the risk component to identify the boundaries of the risk.

Risk evaluation: See Risk assessment.

Risk event status: A measure of the position of a risk event.

Risk Event: A discrete occurrence that may impact the project for better or worse.

Risk exposure: The likely loss or consequence of risk, the combined probability and impact of a risk that is expressed as the result of its probability x impact.

Risk factor: The risk event, probability, or amount at stake.

Risk identification: Part of the risk management plan that involves determining which risk events are likely to affect the project.

Risk Impact: See Impact

Risk management plan: Part of the overall project management plan that includes identifying, analyzing, responding to project risk, monitoring project risk, and closing project risks.

Risk management: A systematic application of policies, procedures, methods, and practices to identifying, analyzing, evaluating, treating, and monitoring risk.

Risk matrix: A matrix with risks located in rows and with impact and likelihood in columns.

Risk mitigation: Actions taken to eliminate or reduce risk by reducing the probability and or impact of occurrence.

Risk Mitigation: Actions taken to eliminate or reduce risk by reducing the probability and or impact of occurrence.

Risk prioritizing: The process of organizing project risks in accordance with their value.

Risk probability: The assessment of the likelihood that a risk event will occur.

Risk quantification: Assessing the probability of risk event occurrence and effect.

Risk ranking: Allocating a classification to the impact and likelihood of a risk.

Risk reduction: Action taken to reduce the likelihood and impact of a risk.

Risk register: A formal record of identified risks listing all the risks identified for the project (For example, the nature of each risk and recording information relevant to its assessment and management).

Risk response control: The process of implementing risk strategies and responding to changes in risk during the life of the project.

Risk response development: Development of strategies and actions to maximize the positive impact of risk events and minimize the negative impact of risk events on a project.

Risk response: Part of the risk management plan that involves the actions required to address the occurrence of a risk event. Contingency plans are collections of risk responses.

Risk sharing: A reduction of risk by sharing it with others.

Risk summary: The description of each risk factor (For example, effect, ownership, and recommendations for response).

Risk symptom: Signs, warnings, or indications of an actual risk event (For example, poor project team morale is an early warning signal of a schedule delay or cost overruns. Also called risk trigger.

Risk transfer: A formal agreement between two parties for the transfer and acceptance of a deliverable where the costs of risk are transferred from one party to the other.

Risk treatment: An implementation of an agreed action and its implementation to deal with an identified risk.

Risk trigger: An event or condition that causes a risk to occur. A trigger is a root cause of such events

Risk: The likelihood of the occurrence of an event expressed in terms of probability, impact, and a triggering event. See also Project risk

Roadblock: An obstacle to project progress (For example, aa barrier that prevents the project from performing project activities).

Robust design: An appropriate design suitable for performing its intended function under different circumstances.

ROI: See Return on investment.

Rollout: Phased introduction of a project's product or service into an organization.

ROS: See Return on sales.

Royalty: A payment made to a party for the use of its intellectual assets (For example, a percentage of profit or other criteria agreed to by the parties involved).

S

Safety plan: The standards and methods intended to minimize the likelihood of accident or damage to people or equipment.

Sample frequency: The ratio of the number of randomly selected units of product for inspection to those passing an inspection.

Sample plan: The formal documentation of sample size, acceptance, and rejection criteria.

Sample size: The number of units of product in the sample selected for inspection.

Sample unit: The unit of the product selected to be part of a sample.

Sample: The units of the product selected from a batch at random or according to an established sampling plan for quality control purposes. See also Representative sample and Biased sample.

Scalable model: A model that generates results for different sizes of projects.

Schedule analysis: See Network analysis.

Schedule baseline: The approved project schedule that serves as the basis for measuring and reporting schedule performance.

Schedule Compression: See duration compression.

Schedule control: Controlling changes and modifications to the project schedule.

Schedule dates: The start and finish dates calculated with regard to resource calendars and constraints.

Schedule development: Analyzing activity sequences, activity durations, and resource requirements to create the project schedule.

Schedule management plan: Part of the overall project management plan, a document used to define scheduling resources, monitor, control, and management of schedule changes.

Schedule management: The process involving the identification and definition of the project schedule that must describe project activates sequence, duration, and dependencies.

Schedule performance index (SPI): The ratio of work performed compared to work planned (For example, if SPI is less than 1, the project is behind schedule). See also Earned value.

Schedule refinement: See schedule update

Schedule revision: See Schedule update

Schedule risk: A risk that jeopardizes completing the project according to the approved schedule.

Schedule simulation: A simulated model of project scheduled alternatives used to quantify the risks of various schedule alternatives (For example, schedule simulations are based on Monte Carlo analysis).

Schedule update: A schedule modification to reflect the current status of the project.

Schedule variance (SV): The difference between the scheduled completion of an activity and the actual completion

Schedule work unit: A calendar time unit when work may be performed on an activity.

Schedule: A time-sequenced project timeline, identifying project activities start, complete dates, and resources required.

Scheduled finish date (SF): A point in time that activity was scheduled to finish (For example, an SF is generally within the range of dates bordered by the activity early finish date and the activity late finish date).

Scheduled start date (SS): A point in time that activity was scheduled to start (For example, The SS is generally within the range of dates bordered by the activity early start date and the activity late start date).

Scheduled start: The earliest date on which an activity can start taking into consideration resource availability and constraints.

Scheduling: The process used to determine the overall project duration and when activities and events are planned to be performed.

Scope baseline approval: A formal approval of the scope baseline by the project stakeholders.

Scope Baseline: See baseline.

Scope change control system: A procedure that is used to change the scope of the project (For example, change documentation, evaluation, and approval procedures).

Scope change control: Also called scope change management. The process of ensuring that all changes to the project scope are evaluated, and their implications to the project plan are considered in making a decision to make the change, postpone it, or reject it.

Scope change: Any change in a project scope that requires a change in the project's cost or schedule.

Scope constraint: Limitations affecting project scope.

Scope cost: An estimate of the cost of the work as defined in the project scope statement.

Scope creep: The unconscious progressive increase of the project scope resulting from uncontrolled changes to requirements.

Scope definition: Breaking down the major deliverables into smaller, more manageable components to provide better control.

Scope management plan: Part of the overall project management that includes the processes of making sure that the project contains all, and only the work required, to accomplish the project successfully. It consists of initiation, scope planning, scope definition, scope verification, and scope change control.

Scope management: The process by which the deliverables and work to produce them are identified and defined. Identification and definition of the scope must describe what the project will include and what it will not include.

Scope of work: A description of the work to be accomplished to satisfy the objectives of the project.

Scope Planning: The practice of developing a written scope statement of the principle project deliverables with the project's justification (business case) and objectives.

Scope risks: A risk that jeopardizes completing the project according to the approved scope.

Scope statement: A document describing the products or services to be delivered. The scope statement details the project deliverables and describes the primary objectives. The objectives should include measurable success criteria for the project.

Scope Verification: Ensuring the satisfactory completion of all identified project deliverables.

Scope: The sum of the products and services to be delivered as a project.

Screening inspection: An inspection where items inspected and found defective will be removed based on a particular criterion.

Screening: A review to analyze and select the alternatives for the proposed solution.

S-Curve: A graphic display of cumulative costs, labor hours, or other quantities, plotted against time. Shaped like an S (For example, flatter at the beginning and end, steeper in the middle, which is typical of most starts slowly, accelerates, and then tails off).

Sealed bidding: A requirement to submit competitive bids in a closed sealed envelope (for example, submission to public tender followed by the public opening of the envelops).

Second source: An alternative source of service supply to encourage competition.

Secondary failure: A system failure that may be caused by excessive use of a product.

Secondary float: See Free float.

Secondary risk: A risk that can occur as a result of treating risk.

Secondary risk: See Residual risk

Secretive management style: A management style where the project manager is not open to the harm of the project (For example, Secretive managers are very insecure and paranoid about losing their jobs. Because of their constant fear, they micromanage teams and keep things to themselves.

Self-inspection: A process where the individual performing a project activity also conducts the examination to ensure that requirements are met.

Sensitivity analysis: The assessment of how the uncertainty of the impact a change has on the outcome of a project.

Sequence: The order in which activities will happen with respect to each other. Successor and predecessor relationships are developed in a network format to allow the project team to have visibility of the activities of the workflow.

Sequencing tasks: A scheduling process where tasks are positioned in series or in parallel to one another based on dependencies between them.

Service contract: An exclusive contract that involves the time and effort of a contractor to perform project tasks rather than provide a physical product.

Severance pay: A financial compensation on top of a regular salary as a result of involuntary termination.

Shared leadership style: A management style where the project manager shares leadership responsibilities with other team members

Should cost Estimates: An estimate of the cost of a product or service used to provide an assessment of the fairness of the contractor's proposed cost.

Significant variance: A difference between the plan and actual performance that jeopardizes the project objectives.

Simulation: An emulation of a process usually conducted a number of times to better understand the process in order to measure its outcomes under different circumstances.

Single point of contact (SPOC): See Point of contact.

Single source: A sole supplier of the entire quantity of goods or services for a project.

Situation analysis: A review intended to identify realities and variables that may influence a situation.

Six sigmas: A set of techniques and tools for process improvement. An orderly, data-driven approach and methodology for eliminating defects (For example, driving toward six standard deviations from the mean to the nearest specification's limit) in any process – from manufacturing to transactional and from product to service).

Slack: A calculated time span where an event has to occur within the logical and imposed constraints of the network, without affecting the total project duration.

Slip chart: A representation of the predicted completion dates of milestones, also referred to as trend chart.

Slippage: The amount of slack or float used by the current activity due to a delayed start or increased duration.

SMART: represents the elements for a well-worded objective, namely Specific, Measurable, Achievable, Realistic, Time-Bound.

Smoothing: See conflict management.

Soft logic: See Discretionary dependency.

Soft project: A project that is intended to produce change but does not have a physical end product.

Soft skills: A combination of people skills (For example, social skills, communication skills, character or personality traits, attitudes, career attributes, social intelligence, and emotional intelligence, conflict management, and negotiation).

Software: A set of computer instructions to carry out various applications and tasks.

Solicitation planning: Documenting project requirements and identifying potential sources.

Solicitation: Obtaining quotations, bids, offers, or proposals as appropriate.

SOR: See Statement of requirements.

Source selection: Choosing from among potential contractors.

Sources of risk: Classes of possible risk events that may affect the project positively or negatively (For example, technical risk, customer risk, or delivery risk).

SOW: See Statement of work.

Span of control: States the number of subordinates under the project manager's direct control. (For example, a project manager with five direct reports has a span of control of five).

Special cause variation: A shift in output caused by a specific factor such as environmental conditions or process input parameters. It is a measure of process control. See also Common cause variation.

Specialist: An expert in a particular in an organization.

Specifications: A detailed statement of project deliverables that result from requirements definition and design. Specifications (For example, specifications are the basis for acceptance criteria used in scope verification and quality control).

SPI: See schedule performance index.

Spiral model: A risk-driven software development process model that is based on the unique risk patterns of a given project. The model guides a team to adopt elements of one or more process models (For example, incremental, waterfall, or evolutionary prototyping).

Stakeholder analysis: An assessment of project stakeholder requirements information and communication needs.

Stakeholder management plan: Part of the overall project management plan includes the process of identification, needs and requirements, analysis, and planning to communicate, negotiate, and influence stakeholders.

Stakeholder management: A systematic identification, analysis, and planning to communicate, negotiate, and influence stakeholders.

Stakeholders value analysis: A process of identifying all project stakeholders, grouping them according to their levels of project participation, interest, or influence, and determining how best to involve and communicate each of these stakeholder groups throughout the project.

Stakeholders: See Project stakeholders.

Standard cost: A pre-computed and established cost incurred on a project.

Standard operating procedure (SOP): A comprehensive step by step procedure for a particular operation.

Standard procedure: A formal mechanism of performing a specific type of work wherever it is performed.

Standard wage rate: A typical base salary of an employee before any additions are computed.

Standard: An established, documented measure to prescribe a specific consensus solution by an authority as a regulation for the measure of quantity or value, to a repetitive design, operations, or maintenance situation.

Start Date: The date project work is officially scheduled to start.

Start to finish: Activity relationship in a precedence diagramming network where an activity must start before the successor activity can finish. See logical relationship.

Start to start lag: The minimum amount of wait time between the start of one activity and the start of its successor(s).

Start to start: activity relationship in a precedence diagramming network where an activity must start before the successor activity can start.

Starting activity of a project: An activity with no preceding activities but has a succeeding one or more activities.

Statement of requirements (SOR): A document stating a business problem, where the intended procured items are offered as a problem to be solved, rather than a clearly specified product or service.

Statement of work (SOW): A documented description of products or services to be supplied under the contract.

Statistical sampling: Part of a population of interest for examination.

Statistical sums: Total project costs based on the cost of individual work

Status date: A specified date to which current progress is reported. Also called Time now

Status report: A report describing the current status of the project and distributed to all stakeholders (For example, the status of performance, issues, risks, and actions following a project status meeting).

Status review meeting: Regularly scheduled meetings held to review the status of the project.

Steering committee: A group of stakeholders assigned to monitor the project and give guidance to the project manager.

Steering group: See Steering committee.

Stop-work order: A formal request to halt any ongoing project work (For example, due to overwhelming circumstances, rules, nonconformance, or funding).

Storming: This is the necessary but painful second stage. At this point, the festive period is over, and team members are more likely to try to express their own individuality and resist group pressures. Conflicts resulting from different personalities, styles, and roles may arise, making it difficult for the team to work effectively together. Some re-planning may be needed following a false start. The critical concern becomes "who's in charge?". It can get rough in the storming stage, critical approaches in this stage include group problem solving, reconciling differences, relieving tension, putting things in the right context, and making compromises.

Straight-line method of depreciation: A standard method of depreciation where the value of a fixed asset is reduced gradually over its useful life.

Strategic partnership: A business relationship between independent organizations in support of a mutual strategy.

Strategic plan: A type of plan that is tightly linked to the organization's strategy, mission, vision, and objectives.

Strategy: A high-level plan to achieve one or more goals.

Strengths weaknesses opportunities threats (SWOT) analysis: A strategic planning technique used to help a project or organization identify strengths, weaknesses, opportunities, and threats related to project planning.

Strong matrix: An organizational structure where the project manager assumes more authority over the project of human resources than the functional manager. See also Weak matrix.

Structured walkthrough: A comprehensive review of the project requirements, design, and implementation of a project conducted by an expert team.

Subgrant: A grant made by one organization using funds previously granted to it by another.

Subgrantee: An entity to which a subgrant is awarded and which is accountable to the Grantee for the use of the funds provided.

Sub optimization: The practice of focusing on one component of a project, making changes intended to improve that specific component while ignoring the effects changes may have on the other components.

Subcontract: A contractual agreement to transfers the responsibility and effort of providing goods or services from one organization to another.

Subcontracted items: Project products or services provided or performed by an entity other than the prime contractor.

Subcontracting process: The process of acquiring resources, products, or services from external sources.

Subcontractor: A group or individual providing products or services to the project. Commonly, sub-contractors are considered to be vendors

Subject matter expert (SME): An expert in a certain part of the project's content anticipated to offer input to the project team regarding business, scientific, engineering, or other subjects.

Subnet: A section of a project network diagram in lieu of a subproject.

Subnetwork: See Subnet.

Subproject: A subproject is a group of activities forming a smaller project, which in turn is a part of a larger project.

Subtask: A breakdown of a task into smaller work elements.

Successor activity: An activity whose start or finish depends on the start or finish of predecessor activity.

Summary schedule: See milestone schedule and Master schedule.

Sunk costs: An unavoidable cost that has already been incurred and cannot be recovered.

Supercritical activity: An activity that is behind in schedule resulting in a negative float value.

Supplementary agreement: Mutually agreed on changes by involved parties to accompany a central contract (For example, amendments and modifications).

Supplementary conditions: Changes to standard conditions developed for goods or services (For example, modifications or deletions).

Supplementary information: Any added information collected from additional sources.

Supplier ranking: See Procurement ranking

Supplier: A person or organization responsible for the performance of a contract or subcontract.

Surety: A legally liable person who takes responsibility for another's performance (For example, their appearing in court or paying a debt).

SV: Schedule variance

System architecture: The conceptual model that defines the structure, behavior of a system. A formal description and illustration of a system, organized in a way that supports reasoning about the structures and behaviors of the system

System design: The process of defining the structure, interfaces, and data for a system to satisfy specified requirements. Systems design could be seen as the application of systems theory to product development

System engineering: An interdisciplinary field of engineering and engineering management that focuses on how to design and manage complex systems over their life cycles. At its core, systems engineering utilizes systems thinking principles to organize this body of knowledge

System flowchart: See Process flowchart.

System interfaces: Describes how the systems or parts of a system communicate by allowing information to flow between them.

System life: The span of time begins when an information technology application is installed and ends when the users' need for it disappears.

System: A methodical assembly of activities or parts forming a logical and connected scheme or unit, the entire technical output of the project.

Systems and procedures: A description of the standard methods, practices, and procedures of handling frequently occurring events within the project.

Systems approach: An style in the management field that stresses the interactive nature and interdependence of external and internal factors in an organization (For example, a systems approach is commonly used to evaluate market elements which affect the profitability of a business.

Systems management: Management that includes the main activities of systems analysis, design and engineering, and development.

Systems thinking: A holistic approach to analysis that focuses on the way that a system's constituent parts interrelate and how systems work overtime and within the context of larger systems.

T

Taguchi method: Statistical methods developed to improve the quality of manufactured goods. (For example, also applied to engineering, biotechnology, marketing, and advertising and sometimes called robust design methods).

Tangible capital asset: See Asset and Physical assets.

Target completion date: See Completion date.

Target cost: See Contract target cost and Planned cost.

Target date: A date imposed on an activity or project by the user. There are two types of target dates; target start dates, and target finish dates.

Target finish date (TF): The date a project or an activity is targeted to finish.

Target schedule: See Baseline.

Target start date (TF): The date a project or an activity is targeted to start.

Task Dependency: See Dependency

Taskforce: A team of skilled resources charged with a specific task (for example, solve a specific project problem or develop a particular procedure).

Task type: A unique identification of a task (For example, by resource requirement, skill, or any other characteristic).

Task: A well-defined component of project work.

Team building: the ability to gather the right people to join a project team and get them working together for the benefit of a project. Understanding the stages of growth that teams typically undergo can assist the project leader in predicting and influencing the performance capabilities of a team (For example, recognizing that all teams go through stages of Forming, Storming, Norming, Performing and Adjourning) See, Forming, Storming, Norming, Performing and Adjourning.

Team development: developing skills, as a group and individually, that enhance project performance.

Team: See Project team.

Teamwork: collaborative work towards a common goal as distinct from other ways individuals may work within a group.

Technical assurance: The monitoring of the technical integrity of products.

Technical authority: See expert authority.

Technical baseline: The project's work breakdown structure.

Technical guide: A document that guides the project manager and project team on planning the construction of products.

Technical interfaces: Formal and informal working and reporting relationships among different technical disciplines involved in a project.

Technical products: Products designed by a project for an end-user.

Technical project leader: Person who serves primarily as the senior technical consultant on a team.

Technical quality administration: The process of developing a plan to monitor and control the technical aspects of the project to ensure its satisfactory completion.

Technical quality specification: A formal and proper documentation of the specific project requirements (For example, project design, measurement specification required, and other requirements to meet the objectives of the project).

Technical requirements: A detailed description of the features and specifications of the deliverable to provide the project the team with guidance on what needs to be accomplished in the project.

Technical specification: See Specification.

Telecommuting: Performing project work and activities from a remote location (For example, work at home).

Templates: A set of sample guidelines, outlines, forms, or checklists.

Tender: A solicitation of bids or proposals for goods or services. See also Solicitation.

Term contract: A type of cost-plus-fixed-fee contract here the scope of work is specified in terms of the level of effort and duration.

Termination by addition: Ending the project by transferring it into an existing or newly created organizational entity.

Termination by extinction: Ending the project due to a successful or unsuccessful conclusion.

Termination by integration: Ending the project by integrating the project's output with an ongoing operation.

Termination by starvation: Significantly dropping the project's budget so that progress stops without formally ending the project.

Termination manager: An individual responsible for closing out a project.

Termination: completion of the project, either upon formal acceptance of its deliverables by the client and/or the disposal of such deliverables at the end of their life.

Terms and conditions: A reference to all the contract clauses.

Terms of reference: A specification of a team member's responsibilities and authorities within the project.

Theory of constraints (TOC): A methodology for identifying the most critical limiting constraints that stand in the way of achieving a goal and then systematically improving that constraint until it is no longer the limiting factor (For example, bottlenecks in manufacturing).

Theory W management: Approach to software project management in which the project manager tries to make winners of each party involved in the software process.

Theory X management: An authoritarian style where the emphasis is on "productivity, on the concept of a fair day's work, on the evils of feather-bedding and restriction of output, on rewards for performance. It reflects an fundamental belief that management must counteract an inherent human tendency to avoid work.

Theory Y management: An approach to management that assumes that employees are happy to work and will take on additional duties without being forced to.

Theory Z management: An approach to management based upon a combination of American and Japanese management philosophies and characterized by, among other things, long-term job security, consensual decision making, slow evaluation, and promotion procedures, and individual responsibility within a group context.

Three-point estimate: A technique used to generate three estimates for the construction of an approximate probability distribution representing the outcome events (For example, optimistic time estimate, most likely time estimate, and pessimistic time estimate. Three estimates are derived for each activity.

Tied activities: Project activities that have to be performed sequentially or within a predetermined time of each other.

Tiger team: A group of subject matter experts organized by management to evaluate, analyze, and make recommendations for resolving problems.

Tight matrix: A collocated project team, or a unified physical location of the project team.

Time analysis: The process of calculating the activities early and late dates on a project based on the duration of each activity and the logical relations between activities.

Time and materials arrangement: A standard phrase in a contract for construction, product development, or any other services in which the employer agrees to pay the contractor based upon the time spent.

Time-based network: A linked bar chart that shows the logical and time relationships between activities.

Time-limited resource scheduling: Project scheduled dates in which resource constraints may be relaxed to avoid any delay in project completion.

Time-limited scheduling: The process of scheduling activities, so that imposed dates and activities durations are not exceeded.

Time management: A subset of project management, including the processes required to ensure timely completion of the project. It consists of activity definition, activity sequencing, activity duration estimating, schedule development, and schedule control.

Time recording: The recording of effort used on each activity to update the project plan.

Time scaled logic drawing: A displays of the project activities logical connection in the context of a timescale in where horizontal position represents a point in time.

Time scaled network diagram: A project network diagram that shows the positioning of the activity represents the schedule (For example, a bar chart that includes network logic).

Timesheet: A tool used for recording the actual effort expended against the project and non-project activities.

Time value of money: A reference to the more significant benefit of receiving money now rather than an identical sum later.

Time variance: Scheduled time for the work completed less than the actual time.

Tolerance: A range in which a result of a test is deemed acceptable (For example, a range of values above and below the estimated project cost).

Tools and techniques: Devices, methods, and instruments applied to the input to create an output (For example, activities, services, or materials that enable the project team to develop and complete project deliverables).

Top-down estimating: Cost estimating that begins with the top level of the WBS and then works down to successively lower levels. Also called analogous estimating. See Analogous estimate.

Total allocated budget: The sum of all budgets allocated to the project.

Total certainty: A situation where all information is available about a project or an activity.

Total cost: The sum of direct and indirect costs incurred by the project.

Total float: See Float.

Total quality management (TQM): A common approach to implementing a quality improvement program within an organization. Strategic, integrated management systems for customer satisfaction that guides all employees in every aspect of their work.

TQM: Total quality management.

Traceability: The capability to trace project history, it is interpreted as the ability to verify the history, location, or application of an item employing documented recorded identification

Trademark: A symbol or words legally registered or established by use as representing an organization or product (For example, any symbol, word or name or any combination, used or intended to be used to identify and distinguish the goods/services of one seller or provider from those of others, and to indicate the source of the goods/services**).**

Trade-off: An exchange where you give up one thing to get something else in return (For example, accepting a higher project cost in return for more functionalities).

Training: Activities designed to enhance and improve the skills, knowledge, and capabilities of the project team.

Transformational leadership: A of leadership where a leader works with teams to identify needed change, creating a vision to guide the change through inspiration, and executing the change in tandem with committed members of a group.

Transit time: A dependency link that requires time and no other resources.

Tree search: An examination of a number of alternatives that logically branch from each other in the form of a tree (For example, A tree search starts at the root and explores nodes from there, looking for one particular node that satisfies the conditions mentioned in the problem).

Trend analysis: A technique that attempts to predict future events based on recently observed trend data is based on the idea that what has happened in the past gives an idea of what will happen in the future.

Trend chart: See Slip chart

Triple constraint: A reference to the generally regarded as the three most important factors that a project manager should to consider in any project time, cost, and scope.

TS: Target start date

Turnaround Report: A report created to enter progress status against a list of activities that are scheduled to be in progress during a particular time window.

U

Unacceptable risk: A risk that is significant enough to jeopardize the project (For example, significant enough to jeopardize an organization's strategy or human lives).

Unallowable cost: The cost that is incurred by a contractor which is not chargeable to a project.

Uncertainty: A situation where only part of the information is available to decide.

Unit price contracts: A contract where the contractor is paid an agreed amount per unit of and the total value of the contract is the sum of the quantities needed to complete the work (For example, $100 per hour).

Unlimited schedule: An infinite schedule that is produced without resource constraint.

Unmanageable risk: A project risk that is practically impossible to reduce its likelihood of occurrence or amount at stake.

Unpriced changes: Contract changes that are authorized and agreed upon, but the cost of which are negotiated during execution.

Unsolicited proposal: A written proposal that is not a response to a formal request for a proposal but submitted for the purpose of obtaining a contract.

Update: The latest and current information on a project.

Useful life: The total time a product or a system will provide value to its user.

User-friendly: The degree of learning ease and use of a computer software package or application.

User requirements: The business need that the project is intended to meet.

Users: The group of people who are intended to benefit from the project.

Utility theory: A reference to the satisfaction that each choice provides to the decision-maker. With the assumption that decision is made based on the maximization principle, according to which the best choice is the one that provides the highest satisfaction (For example, a willingness to take a risk in light of the levels of the reward).

V

Validation: the exercise of examining or proving the validity or accuracy, Evaluation of a product against its specified requirements.

Value analysis: An approach to improving the value of a product or process by understanding its constituent components and their associated costs and find improvements to the components by either reducing their cost or increasing the value of the functions.

Value chain: a set of activities that a firm operating in a specific industry performs to deliver a valuable product for the market

Value engineering: A technique for analyzing qualitative and quantitative costs and benefits of parts of a proposed system.

Value management: A planned means of improving business effectiveness that includes the use of management techniques.

Value planning: A method of assessing the desirability of a proposal based on the value that will accrue to the organization from that proposal before significant investment is made.

Value: A standard, principle, or quality considered worthwhile or desirable.

Variable cost: Costs that change as the quantity of the good or service that a business produces changes. Variable costs are the sum of marginal costs over all units produced. They can also be considered standard costs (for example, fixed costs and variable costs make up the two components of total cost**).**

Variable sampling Plan: An acceptance sampling technique. Plans for variables are intended for quality characteristics that are measured on a continuous scale.

Variance analysis: A comparative analysis of actual project results to planned or expected results.

Variance at completion: The difference between budget at completion and estimate at completion.

Variance report: A formal report of project performance involves documentation difference in cost and schedule between actuals and planned.

Variance: A schedule or cost difference between the actual and planned values on a project.

Variation orders: A formal authorization for a change or variation on a project.

Vendor: An organization or individuals providing products or services under contract to the project performance group.

Vendors conference: See bidders conference

Vendors conference: See Bidders conference.

Verification: Evaluation of the correctness of a process output based on an agreed specification.

Version: A form of application or system that reflects significant changes in functions or features.

Virtual team: A project team that works together from different geographic locations and relies on communication technology such as email, video, or voice conferencing services to collaborate.

Vision Statement: A formal statement that captures the long-term aspiration of what the organization wants to achieve (For example, A vision statement must be inspirational, memorable, and be recognized for leadership, innovation, and excellence).

W

Waiver: A voluntary relinquishment or surrender of some known right or privilege (For example, a law restricted the lendinglimits, but when financial institutes exceeded the limits, they obtained waivers.

Walk-through: An examination of project requirements, design, or implementation by a qualified third party to ensure that the project objectives will be met.

War room: A designated area for project meetings conferences and serves as a command and control center.

Warranty clause: A specific clause in a contract to provide the buyer with additional time after delivery to correct defects or adjustment.

Warranty: A promise, a written guarantee, issued to the purchaser of a product by its vendor promising to repair or replace it if necessary, within a specified period.

WBS dictionary: A collection of work package descriptions (For example, planning information, schedule dates, cost budgets, and resources).

WBS: See Work breakdown structure.

Weak matrix: An organizational structure where the functional manager assumes more authority over the project human resources than the project manager. See also Strong matrix.

Weighing system: A set of comparative values that permit assigning different levels of importance to each of several items (For example, as contractor qualifications where the relative weights consider the overall significance of a factor, in terms of both its necessary quality and relative importance.

What if simulation: Changing the values of the parameters of the project network to study its behavior under various conditions of its operation.

What-if analysis: The process of evaluating alternative scenarios and strategies.

White paper: An authoritative report or guide of concise information about a complex issue and presents the issuer's philosophy on the matter. It is meant to help readers understand an issue, solve a problem, or make an informed decision.

Win-win: Outcome of conflict management that results in both parties being better off than when the conflict started.

Withdrawing: See conflict management.

Withholding: An amount not paid due to a failure to perform some responsibility under a contract.

Work acceptance: See Customer acceptance

Work authorization system: The formal procedure used to authorize project work (For example, as part of working methodology).

Work authorization: See Authorization

Work breakdown code: A code that represents an element's relationship in a work breakdown structure (For example, the element family tree).

Work breakdown structure (WBS): A hierarchically deliverable structure that represents a group of project elements that defines and categorizes the total scope of the project (For example, in a WBS, each descending level represents an increasingly detailed definition of the project work.)

Work element: See Task.

Work Flow: The relationship between the activities in a project from start to finish taking into consideration all types of activity relationships.

Work Item: See Activity.

Work Load: The number of work units assigned to a resource over a period of time.

Work Package: A group of related tasks that are defined at the same level within a project work breakdown structure at which project accounting is performed (For example, a week in duration and performed by an individual or small workgroup).

Work release: See work authorization.

Work results: Outcome of activities performed to accomplish the project.

Work statement: See Statement of work.

Work Units: Work units provide measurement units for resources. For example, people as a resource can be measured by the number of hours they work.

Work: The total number of hours, people, or effort required to complete a project activity or a task.

Workaround: An unplanned response to an adverse risk event. It is distinguished from the contingency plan in that a workaround is not planned for the occurrence of the risk event.

Working calendar: See Project calendar

Workload: Workload is the number of work units assigned to a resource over a period of time.

Z

Zero defect (ZD): Quality standard that states that no less than 100 percent quality should be the goal of a project.

Zero Float: A condition where there is no excess time between activities. An activity with zero float is considered a critical activity (For example, if the duration of any critical activity is increased the activity slips, the project finish date will slip).

Zero one hundred approach: An earned value method for project activities where no value is earned when an activity starts, but 100 percent of the value is earned when it's completed.

Zero variance: The planned schedule or cost is equal to the actual schedule or cost of an activity or project.

www.ingramcontent.com/pod-product-compliance
Lightning Source LLC
Chambersburg PA
CBHW070534200326
41519CB00013B/3037